John Ruskin

Wayside flowers in England and Scotland

John Ruskin

Wayside flowers in England and Scotland

ISBN/EAN: 9783743345423

Manufactured in Europe, USA, Canada, Australia, Japa

Cover: Foto ©berggeist007 / pixelio.de

Manufactured and distributed by brebook publishing software (www.brebook.com)

John Ruskin

Wayside flowers in England and Scotland

PROSERPINA.

STUDIES OF WAYSIDE FLOWERS,

WHILE THE AIR WAS YET PURE

AMONG THE ALPS, AND IN THE SCOTLAND AND ENGLAND WHICH MY FATHER KNEW.

BY

JOHN RUSKIN, LL.D.,

HONORARY STUDENT OF CHRISTCHURCH, AND HONORARY FELLOW OF CORPUS CHRISTI COLLEGE, OXFORD.

VOL. II.

NEW YORK:

JOHN WILEY & SONS,

15 ASTOR PLACE.

1886.

PROSERPINA.

CHAPTER I.

VIOLA.

1. ALTHOUGH I have not been able in the preceding
volume to complete, in any wise as I desired, the account
of the several parts and actions of plants in general, I
will not delay any longer our entrance on the examina-
tion of particular kinds, though here and there I must
interrupt such special study by recurring to general
principles, or points of wider interest. But the scope of
such larger inquiry will be best seen, and the use of it
best felt, by entering now on specific study.

I begin with the Violet, because the arrangement of
the group to which it belongs—Cytherides—is more ar-
bitrary than that of the rest, and calls for some immediate explanation.

2. I fear that my readers may expect me to write
something very pretty for them about violets: but my
time for writing prettily is long past; and it requires
some watching over myself, I find, to keep me even

from writing querulously.   For while, the older I grow,
very thankfully I recognize more and more the number
of pleasures granted to human eyes in this fair world, I
recognize also an increasing sensitiveness in my temper
to anything that interferes with them; and a grievous
readiness to find fault—always of course submissively,
but very articulately—with whatever Nature seems to
me not to have managed to the best of her power;—as,
for extreme instance, her late arrangements of frost this
spring, destroying all the beauty of the wood sorrels;
nor am I less inclined, looking to her as the greatest of
sculptors and painters, to ask, every time I see a narcissus, why it should be wrapped up in brown paper; and
every time I see a violet, what it wants with a spur?

3. What *any* flower wants with a spur, is indeed the
simplest and hitherto to me unanswerablest form of the
question; nevertheless, when blossoms grow in spires,
and are crowded together, and have to grow partly
downwards, in order to win their share of light and
breeze, one can see some reason for the effort of the
petals to expand upwards and backwards also.   But that
a violet, who has her little stalk to herself, and might
grow straight up, if she pleased, should be pleased to do
nothing of the sort, but quite gratuitously bend her
stalk down at the top, and fasten herself to it by her
waist, as it were,—this is so much more like a girl of the
period's fancy than a violet's, that I never gather one
separately but with renewed astonishment at it.

4. One reason indeed there is, which I never thought of until this moment! a piece of stupidity which I can only pardon myself in, because, as it has chanced, I have studied violets most in gardens, not in their wild haunts, —partly thinking their Athenian honour was as a garden flower; and partly being always led away from them, among the hills, by flowers which I could see nowhere else. With all excuse I can furbish up, however, it is shameful that the truth of the matter never struck me before, or at least this bit of the truth—as follows.

5. The Greeks, and Milton, alike speak of violets as growing in meadows (or dales). But the Greeks did so because they could not fancy any delight except in meadows; and Milton, because he wanted a rhyme to nightingale — and, after all, was London bred. But Viola's beloved knew where violets grew in Illyria,—and grow everywhere else also, when they can,—on a *bank*, facing the south.

Just as distinctly as the daisy and buttercup are *meadow* flowers, the violet is a *bank* flower, and would fain grow always on a steep slope, towards the sun. And it is so poised on its stem that it shows, when growing on a slope, the full space and opening of its flower, —not at all, in any strain of modesty, hiding *itself*, though it may easily be, by grass or mossy stone, 'half hidden,'—but, to the full, showing itself, and intending to be lovely and luminous, as fragrant, to the uttermost of its soft power.

Nor merely in its oblique setting on the stalk, but in the reversion of its two upper petals, the flower shows this purpose of being fully seen. (For a flower that *does* hide itself, take a lily of the valley, or the bell of a grape hyacinth, or a cyclamen.) But respecting this matter of petal-reversion, we must now farther state two or three general principles.

6. A perfect or pure flower, as a rose, oxalis, or campanula, is always composed of an unbroken whorl, or corolla, in the form of a disk, cup, bell, or, if it draw together again at the lips, a narrow-necked vase. This cup, bell, or vase, is divided into similar petals, (or segments, which are petals carefully joined,) varying in number from three to eight, and enclosed by a calyx whose sepals are symmetrical also.

An imperfect, or, as I am inclined rather to call it, an 'injured' flower, is one in which some of the petals have inferior office and position, and are either degraded, for the benefit of others, or expanded and honoured at the cost of others.

Of this process, the first and simplest condition is the reversal of the upper petals and elongation of the lower ones, in blossoms set on the side of a clustered stalk. When the change is simply and directly dependent on their position in the cluster, as in Aurora Regina,* modifying every bell just in proportion as it declines from the perfected central one, some of the loveliest groups of

* Vol. i., p. 212, note.

form are produced which can be seen in any inferior
organism : but when the irregularity becomes fixed, and
the flower is always to the same extent distorted, whatever its position in the cluster, the plant is to be rightly
thought of as reduced to a lower rank in creation.

7. It is to be observed, also, that these inferior forms
of flower have always the appearance of being produced
by some kind of mischief—blight, bite, or ill-breeding;
they never suggest the idea of improving themselves, now,
into anything better; one is only afraid of their tearing
or puffing themselves into something worse. Nay, even
the quite natural and simple conditions of inferior vegetable do not in the least suggest, to the unbitten or un-
blighted human intellect, the notion of development into
anything other than their like : one does not expect a
mushroom to translate itself into a pineapple, nor a
betony to moralize itself into a lily, nor a snapdragon to
soften himself into a lilac.

8. It is very possible, indeed, that the recent phrenzy'
for the investigation of digestive and reproductive oper-
ations in plants may by this time have furnished the
microscopic malice of botanists with providentially dis-
gusting reasons, or demoniacally nasty necessities, for
every possible spur, spike, jag, sting, rent, blotch, flaw,
freckle, filth, or venom, which can be detected in the
construction, or distilled from the dissolution, of vegetable organism. But with these obscene processes and
prurient apparitions the gentle and happy scholar of

flowers has nothing whatever to do. I am amazed and
saddened, more than I can care to say, by finding how
much that is abominable may be discovered by an ill-
taught curiosity, in the purest things that earth is allowed
to produce for us;—perhaps if we were less reprobate in
our own ways, the grass which is our type might con-
duct itself better, even though *it* has no hope but of
being cast into the oven; in the meantime, healthy
human eyes and thoughts are to be set on the lovely
laws of its growth and habitation, and not on the mean
mysteries of its birth.

9. I relieve, therefore, our presently inquiring souls
from any farther care as to the reason for a violet's spur,
—or for the extremely ugly arrangements of its stamens
and style, invisible unless by vexatious and vicious peep-
ing. You are to think of a violet only in its green
leaves, and purple or golden petals;—you are to know
the varieties of form in both, proper to common species;
and in what kind of places they all most fondly live, and
most deeply glow.

" And the recreation of the minde which is taken
heereby cannot be but verie good and honest, for they
admonish and stir up a man to that which is comely and
honest. For flowers, through their beautie, varietie of
colour, and exquisite forme, do bring to a liberall and
gentle manly minde the remembrance of honestie, come-
liness, and all kinds of vertues. For it would be an un-
seemely and filthie thing, as a certain wise man saith, for

him that doth looke upon and handle faire and beautiful
things, and who frequenteth and is conversant in faire
and beautiful places, to have his mind not faire, but
filthie and deformed."

10. Thus Gerarde, in the close of his introductory
notice of the violet,—speaking of things, (honesty, come-
liness, and the like,) scarcely now recognized as desirable
in the realm of England; but having previously observed that violets are useful for the making of garlands
for the head, and posies to smell to;—in which last function I observe they are still pleasing to the British public: and I found the children here, only the other day,
munching a confection of candied violet leaves. What
pleasure the flower can still give us, uncandied, and un-
bound, but in its own place and life, I will try to trace
through some of its constant laws.

11. And first, let us be clear that the native colour of
the violet *is* violet; and that the white and yellow kinds,
though pretty in their place and way, are not to be
thought of in generally meditating the flower's quality
or power. A white violet is to black ones what a black
man is to white ones; and the yellow varieties are, I
believe, properly pansies, and belong also to wild districts for the most part; but the true violet, which I
have just now called ' black,' with Gerarde, " the blacke
or purple violet, hath a great prerogative above others,"
and all the nobler species of the pansy itself are of full
purple, inclining, however, in the ordinary wild violet to

blue. In the 'Laws of Fésole,' chap. vii., §§ 20, 21, I
have made this dark pansy the representative of purple
pure; the viola odorata, of the link between that full
purple and blue; and the heath-blossom of the link between that full purple and red. The reader will do well,
as much as may be possible to him, to associate his study
of botany, as indeed all other studies of visible things,
with that of painting: but he must remember that he
cannot know what violet colour really is, unless he watch
the flower in its *early* growth. It becomes dim in age,
and dark when it is gathered—at least, when it is tied in
bunches;—but I am under the impression that the colour
actually deadens also,—at all events, no other single
flower of the same quiet colour lights up the ground
near it as a violet will. The bright hounds-tongue looks
merely like a spot of bright paint; but a young violet
glows like painted glass.

12. Which, when you have once well noticed, the two
lines of Milton and Shakspeare which seem opposed,
will both become clear to you. The said lines are
dragged from hand to hand along their pages of pilfered
quotations by the hack botanists,—who probably never
saw *them*, nor anything else, *in* Shakspeare or Milton in
their lives,—till even in reading them where they rightly
come, you can scarcely recover their fresh meaning: but
none of the botanists ever think of asking why Perdita
calls the violet 'dim,' and Milton 'glowing.'

Perdita, indeed, calls it dim, at that moment, in think-

ing of her own love, and the hidden passion of it, un-
speakable; nor is Milton without some purpose of using
it as an emblem of love, mourning,—but, in both cases,
the subdued and quiet hue of the flower as an actual tint
of colour, and the strange force and life of it as a part
of light, are felt to their uttermost.

And observe, also, that both of the poets contrast the
violet, in its softness, with the intense marking of the
pansy. Milton makes the opposition directly—

> " the pansy, freaked with jet,
> The glowing violet."

Shakspeare shows yet stronger sense of the difference, in
the "purple with Love's wound" of the pansy, while
the violet is sweet with Love's hidden life, and sweeter
than the lids of Juno's eyes.

Whereupon, we may perhaps consider with ourselves
a little, what the difference *is* between a violet and a
pansy?

13. Is, I say, and was, and is to come,—in spite of
florists, who try to make pansies round, instead of pentagonal; and of the wise classifying people, who say that
violets and pansies are the same thing—and that neither
of them are of much interest! As, for instance, Dr.
Lindley in his 'Ladies' Botany.'

" Violets—sweet Violets, and Pansies, or Heartsease,
represent a small family, with the structure of which
you should be familiar; more, however, for the sake of

its singularity than for its extent or importance, for the
family is a very small one, and there are but few species
belonging to it in which much interest is taken. As
the parts of the Heartsease are larger than those of the
Violet, let us select the former in preference for the
subject of our study." Whereupon we plunge instantly
into the usual account of things with horns and tails.
" The stamens are five in number—two of them, which
are in front of the others, are hidden within the horn of
the front petal," etc., etc., etc. (Note in passing, by the
'*horn of the front*' petal he means the '*spur of the bottom*' one, which indeed does stand in front of the rest,
—but if therefore *it* is to be called the *front* petal—
which is the back one?) You may find in the next paragraph description of a " singular conformation," and
the interesting conclusion that " no one has yet discovered for what purpose this singular conformation was
provided." But you will not, in the entire article, find
the least attempt to tell you the difference between a
violet and a pansy!—except in one statement—and *that*
false! " The sweet violet will have no rival among
flowers, if we merely seek for delicate fragrance; but
her sister, the heartsease, who is destitute of all sweetness, far surpasses her in rich dresses and *gaudy*!!!
colours." The heartsease is not without sweetness.
There are sweet pansies scented, and dog pansies unscented—as there are sweet violets scented, and dog
violets unscented. What is the real difference?

14. I turn to another scientific gentleman—*more* scientific in form indeed, Mr. Grindon,—and find, for another interesting phenomenon in the violet, that it sometimes produces flowers without any petals! and in the pansy, that " the flowers turn towards the sun, and when many are open at once, present a droll appearance, looking like a number of faces all on the ' qui vive.' " But nothing of the difference between them, except something about ' stipules,' of which " it is important to observe that the leaves should be taken from the middle of the stem—those above and below being variable."

I observe, however, that Mr. Grindon *has* arranged his violets under the letter A, and his pansies under the letter B, and that something may be really made out of him, with an hour or two's work. I am content, however, at present, with his simplifying assurance that of violet and pansy together, " six species grow wild in Britain—or, as some believe, only four—while the analysts run the number up to fifteen."

15. Next I try Loudon's Cyclopædia, which, through all its 700 pages, is equally silent on the business ; and next, Mr. Baxter's ' British Flowering Plants,' in the index of which I find neither Pansy nor Heartsease, and only the ' Calathian' Violet, (where on earth is Calathia?) which proves, on turning it up, to be a Gentian.

16. At last, I take my Figuier, (but what should I do if I only knew English?) and find this much of clue to the matter :—

" Qu'est ce que c'est que la Pensée ? Cette jolie plante
appartient aussi ou genre Viola, mais à un section de ce
genre.   En effet, dans les Pensées, les pétales supérieurs
et lateraux sont dirigés en haut, l'inférieur seul est
dirigé en bas : et de plus, le stigmate est urcéole, glo-
buleux."

And farther, this general description of the whole
violet tribe, which I translate, that we may have its full
value :—

" The violet is a plant without a stem (tige),—(see
vol. i., p. 154,)—whose height does not surpass one or
two decimetres.   Its leaves, radical, or carried on stolons,
(vol. i., p. 158,) are sharp, or oval, crenulate, or heart-
shape.   Its stipules are oval-acuminate, or lanceolate.
Its flowers, of sweet scent, of a dark violet or a reddish
blue, are carried each on a slender peduncle, which bends
down at the summit.   Such is, for the botanist, the
Violet, of which the poets would give assuredly another
description."

17. Perhaps ; or even the painters ! or even an ordinary unbotanical human creature !   I must set about my
business, at any rate, in my own way, now, as I best can,
looking first at things themselves, and then putting this
and that together, out of these botanical persons, which
they can't put together out of themselves.   And first, I
go down into my kitchen garden, where the path to the
lake has a border of pansies on both sides all the way
down, with clusters of narcissus behind them.   And

pulling up a handful of pansies by the roots, I find them "without stems," indeed, if a stem means a wooden thing; but I should say, for a low-growing flower, quiet lankily and disagreeably stalky! And, thinking over what I remember about wild pansies, I find an impression on my mind of their being rather more stalky, always, than is quite graceful; and, for all their fine flowers, having rather a weedy and littery look, and getting into places where they have no business. See, again, vol. i., chap. vi., § 5.

18. And now, going up into my flower and fruit garden, I find (June 2nd, 1881, half-past six, morning,) among the wild saxifrages, which are allowed to grow wherever they like, and the rock strawberries, and Francescas, which are coaxed to grow wherever there is a bit of rough ground for them, a bunch or two of pale pansies, or violets, I don't know well which, by the flower; but the entire company of them has a ragged, jagged, unpurpose-like look; extremely,—I should say,—demoralizing to all the little plants in their neighbourhood : and on gathering a flower, I find it is a nasty big thing, all of a feeble blue, and with two things like horns, or thorns, sticking out where its ears would be, if the pansy's frequently monkey face were underneath them. Which I find to be two of the leaves of its calyx 'out of place,' and, at all events, for their part, therefore, weedy, and insolent.

19. I perceive, farther, that this disorderly flower is

lifted on a lanky, awkward, springless, and yet stiff
flower-stalk ; which is not round, as a flower-stalk ought
to be, (vol. i., p. 155,) but obstinately square, and fluted,
with projecting edges, like a pillar run thin out of an
iron-foundry for a cheap railway station. I perceive
also that it has set on it, just before turning down to
carry the flower, two little jaggy and indefinable leaves,
—their colour a little more violet than the blossom.

These, and such undeveloping leaves, wherever they
occur, are called ' bracts' by botanists, a good word, from
the Latin ' bractea,' meaning a piece of metal plate, so
thin as to crackle. They seem always a little stiff, like
bad parchment,—born to come to nothing—a sort of in-
finitesimal fairy-lawyer's deed. They ought to have been
in my index at p. 255, under the head of leaves, and are
frequent in flower structure,—never, as far as one can
see, of the smallest use. They are constant, however, in
the flower-stalk of the whole violet tribe.

20. I perceive, farther, that this lanky flower-stalk,
bending a little in a crabbed, broken way, like an obsti-
nate person tired, pushes itself up out of a still more
stubborn, nondescript, hollow angular, dogseared gas-
pipe of a stalk, with a section something like this,

 but no bigger than with a quantity of

ill-made and ill-hemmed leaves on it, of no describable
leaf-cloth or texture,—not cressic, (though the thing does

altogether look a good deal like a quite uneatable old
watercress) ; not salvian, for there's no look of warmth
or comfort in them; not cauline, for there's no juice in
them; not dryad, for there's no strength in them, nor
apparent use : they seem only there, as far as I can make
out, to spoil the flower, and take the good out of my
garden bed. Nobody in the world could draw them,
they are so mixed up together, and crumpled and hacked
about, as if some ill-natured child had snipped them with
blunt scissors, and an ill-natured cow chewed them a
little afterwards and left them, proved for too tough or
too bitter.

21. Having now sufficiently observed, it seems to me,
this incongruous plant, I proceed to ask myself, over it,
M. Figuier's question, ' Qu'est-ce c'est qu'un Pensée ? '
Is this a violet—or a pansy—or a bad imitation of both ?

Whereupon I try if it has any scent : and to my much
surprise, find it has a full and soft one—which I suppose
is what my gardener keeps it for! According to Dr.
Lindley, then, it must be a violet! But according to M.
Figuier,—let me see, do its middle petals bend up, or
down ?

I think I'll go and ask the gardener what *he* calls it.

22. My gardener, on appeal to him, tells me it is the
' Viola Cornuta,' but that he does not know himself if it
is violet or pansy. I take my London again, and find
there were fifty-three species of violets, known in his
days, of which, as it chances, Cornuta is exactly the last.

'Horned violet': I said the green things were *like* horns!—but what is one to say of, or to do to, scientific people, who first call the spur of the violet's petal, horn, and then its calyx points, horns, and never define a 'horn' all the while!

Viola Cornuta, however, let it be; for the name does mean *some*thing, and is not false Latin. But whether violet or pansy, I must look farther to find out.

23. I take the Flora Danica, in which I at least am sure of finding whatever is done at all, done as well as honesty and care can; and look what species of violets it gives.

Nine, in the first ten volumes of it; four in their modern sequel (that I know of,—I have had no time to examine the last issues). Namely, in alphabetical order, with their present Latin, or tentative Latin, names; and in plain English, the senses intended by the hapless scientific people, in such their tentative Latin :—

(1) Viola Arvensis.   Field (Violet) . . . No. 1748
(2)   "   Biflora.   Two-flowered . . .   46
(3)   "   Canina.   Dog   1453
(3B) "   Canina.   Var. Multicaulis (many-
                      stemmed), a very singular sort of
                      violet—if it were so! Its real dif-
                      ference from our dog-violet is in
                      being pale blue, and having a
                      golden centre   2646

(4) Viola Hirta.   Hairy   618

(5) "   Mirabilis.   Marvellous   1045

(6) "   Montana.   Mountain   1329

(7) "   Odorata.   Odorous   309

(8) "   Palustris.   Marshy   83

(9) "   Tricolor.   Three-coloured . . .   623

(9B) "   Tricolor.   Var. Arenaria, Sandy
    Three-coloured.   2647

(10) "   Elatior. Taller   68

(11) "   Epipsila. (Heaven knows what: it is
    Greek, not Latin, and looks as if
    it meant something between a
    bishop and a short letter e) . .   2405

I next run down this list, noting what names we can keep, and what we can't; and what aren't worth keeping, if we could: passing over the varieties, however, for the present, wholly.

(1) Arvensis. Field-violet. Good.

(2) Biflora. A good epithet, but in false Latin. It is to be our Viola aurea, golden pansy.

(3) Canina. Dog. Not pretty, but intelligible, and by common use now classical. Must stay.

(4) Hirta. Late Latin slang for hirsuta, and always used of nasty places or nasty people; it shall not stay. The species shall be our Viola Seclusa,—Monk's violet—meaning the kind of monk who leads a rough life like Elijah's, or the Baptist's,

2

or Esau's—in another kind. This violet is one of
the loveliest that grows.

(5) Mirabilis. Stays so; marvellous enough, truly: not
more so than all violets; but I am very glad to
hear of scientific people capable of admiring any-
· thing.

(6) Montana. Stays so.

(7) Odorata. Not distinctive;—nearly classical, however. It is to be our Viola Regina, else I should
not have altered it.

(8) Palustris. Stays so.

(9) Tricolor. True, but intolerable. The flower is the
queen of the true pansies: to be our Viola Psyche.

(10) Elatior. Only a variety of our already accepted
Cornuta.

(11) The last is, I believe, also only a variety of Palustris. Its leaves, I am informed in the text, are
either " pubescent-reticulate-venose-subreniform,"
or " lato-cordate-repando-crenate ;" and its stipules
are " ovate-acuminate-fimbrio-denticulate." I do
not wish to pursue the inquiry farther.

24. These ten species will include, noting here and
there a local variety, all the forms which are familiar to
us in Northern Europe, except only two ;—these, as it
singularly chances, being the Viola Alpium, noblest of
all the wild pansies in the world, so far as I have seen or
heard of them,—of which, consequently, I find no pic-

ture, nor notice, in any botanical work whatsoever; and the other, the rock-violet of our own Yorkshire hills.

We have therefore, ourselves, finally then, twelve following species to study. I give them now all in their accepted names and proper order,—the reasons for occasional difference between the Latin and English name will be presently given.

(1)	Viola	Regina.	Queen violet.
(2)	"	Psyche.	Ophelia's pansy.
(3)	"	Alpium.	Freneli's pansy.
(4)	"	Aurea.	Golden violet.
(5)	"	Montana.	Mountain Violet.
(6)	"	Mirabilis.	Marvellous violet.
(7)	"	Arvensis.	Field violet.
(8)	"	Palustris.	Marsh violet.
(9)	"	Seclusa.	Monk's violet.
(10)	"	Canina.	Dog violet.
(11)	"	Cornuta.	Cow violet.
(12)	"	Rupestris.	Crag violet.

25. We will try, presently, what is to be found out of useful, or pretty, concerning all these twelve violets; but must first find out how we are to know which are violets indeed, and which, pansies.

Yesterday, after finishing my list, I went out again to examine Viola Cornuta a little closer, and pulled up a full grip of it by the roots, and put it in water in a wash-hand basin, which it filled like a truss of green hay.

Pulling out two or three separate plants, I find each
to consist mainly of a jointed stalk of a kind I have not
yet described,—roughly, some two feet long altogether;
(accurately, one 1 ft. 10½ in.; another, 1 ft. 10 in.; another, 1 ft. 9 in.—but all these measures taken without
straightening, and therefore about an inch short of the
truth), and divided into seven or eight lengths by clumsy
joints where the mangled leafage is knotted on it; but
broken a little out of the way at each joint, like a rheu-
matic elbow that won't come straight, or bend farther;
and—which is the most curious point of all in it—it is
thickest in the middle, like a viper, and gets quite thin
to the root and thin towards the flower; also the lengths
between the joints are longest in the middle: here I
give them in inches, from the root upwards, in a stalk
taken at random.

1st (nearest root)	.	.	.	0¾
2nd	.	.	.	0¾
3rd	.	.	.	1½
4th	.	.	.	1¾
5th	.	.	.	3
6th	.	.	.	4
7th	.	.	.	3¼
8th	.	.	.	3
9th	.	.	.	2¼
10th	.	.	.	1½

1 ft. 9¾ in.

But the thickness of the joints and length of terminal
flower stalk bring the total to two feet and about an inch

over. I dare not pull it straight, or should break it, but it overlaps my two-foot rule considerably, and there are two inches besides of root, which are merely underground stem, very thin and wretched, as the rest of it is merely root above ground, very thick and bloated. (I begin actually to be a little awed at it, as I should be by a green snake—only the snake would be prettier.) The flowers also, I perceive, have not their two horns regularly set *in*, but the five spiky calyx-ends stick out between the petals—sometimes three, sometimes four, it may be all five up and down—and produce variously fanged or forked effects, feebly ophidian or diabolic. On the whole, a plant entirely mismanaging itself,— reprehensible and awkward, with taints of worse than awkwardness; and clearly, no true 'species,' but only a link.* And it really is, as you will find presently, a link in two directions; it is half violet, half pansy, a 'cur' among the Dogs, and a thoughtless thing among the thoughtful. And being so, it is also a link between the entire violet tribe and the Runners—pease, strawberries, and the like, whose glory is in their speed; but a violet has no business whatever to run anywhere, being appointed to stay where it was born, in extremely contented (if not secluded) places. " Half-hidden from the eye?"—no; but desiring attention, or extension, or corpulence, or connection with anybody else's family, still less.

---

* See ' Deucalion,' vol. ii., chap. i., p. 12, § 18.

26. And if, at the time you read this, you can run out and gather a *true* violet, and its leaf, you will find that the flower grows from the very ground, out of a cluster of heart-shaped leaves, becoming here a little rounder, there a little sharper, but on the whole heart-shaped, and that is the proper and essential form of the violet leaf. You will find also that the flower has five petals; and being held down by the bent stalk, two of them bend back and up, as if resisting it; two expand at the sides; and one, the principal, grows downwards, with its attached spur behind. So that the front view of the flower must be *some* modification of this typical arrangement, Fig. M, (for middle form). Now the statement above quoted from Figuier, § 16, means, if he had been able to express himself, that the two lateral petals in the violet are directed downwards, Fig. II. A, and in the pansy upwards, Fig. II. c. And that, in the main, is true, and to be fixed well and clearly in your mind. But in the real orders, one flower passes into the other through all kinds of intermediate positions of petal, and the plurality of species are of the middle type, Fig. II. B.*

FIG. II.

27. Next, if you will gather a real pansy *leaf*, you will find it—not heart-shape in the least, but sharp oval

* I am ashamed to give so rude outlines; but every moment now is valuable to me: careful outline of a dog-violet is given in Plate X.

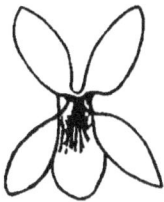

X.

VIOLA CANINA.

Structural Details.

or spear-shape, with two deep cloven lateral flakes at its
springing from the stalk, which, in ordinary aspect, give
the plant the haggled and draggled look I have been
vilifying it for. These, and such as these, "leaflets at
the base of other leaves" (Balfour's Glossary), are called
by botanists 'stipules.' I have not allowed the word
yet, and am doubtful of allowing it, because it entirely
confuses the student's sense of the Latin 'stipula' (see
above, vol. i., chap. viii., § 27) doubly and trebly important in its connection with 'stipulor,' not noticed in
that paragraph, but readable in your large Johnson; we
shall have more to say of it when we come to 'straw'
itself.

28. In the meantime, one *may* think of these things
as stipulations for leaves, not fulfilled, or 'stumps' or
'sumphs' of leaves! But I think I can do better for
them. We have already got the idea of *crested* leaves,
(see vol. i., plate); now, on each side of a knight's crest,
from earliest Etruscan times down to those of the Scalas,
the fashion of armour held, among the nations who
wished to make themselves terrible in aspect, of putting
cut plates or 'bracts' of metal, like dragons' wings, on
each side of the crest. I believe the custom never became Norman or English; it is essentially Greek, Etruscan, or Italian,—the Norman and Dane always wearing
a practical cone (see the coins of Canute), and the Frank
or English knights the severely plain beavered helmet;
the Black Prince's at Canterbury, and Henry V.'s at

Westminster, are kept hitherto by the great fates for us to see. But the Southern knights constantly wore these lateral dragon's wings; and if I can find their special name, it may perhaps be substituted with advantage for 'stipule'; but I have not wit enough by me just now to invent a term.

29. Whatever we call them, the things themselves are, throughout all the species of violets, developed in the running and weedy varieties, and much subdued in the beautiful ones; and generally the pansies have them large, with spear-shaped central leaves; and the violets small, with heart-shaped leaves, for more effective decoration of the ground. I now note the characters of each species in their above given order.

30. I. Viola Regina. Queen Violet. Sweet Violet. 'Viola Odorata,' L., Flora Danica, and Sowerby. The latter draws it with golden centre and white base of lower petal; the Flora Danica, all purple. It is sometimes altogether white. It is seen most perfectly for setting off its colour, in group with primrose,—and most luxuriantly, so far as I know, in hollows of the Savoy limestones, associated with the pervenche, which embroiders and illumines them all over. I believe it is the earliest of its race, sometimes called 'Martia,' March violet. In Greece and South Italy even a flower of the winter.

> " The Spring is come, the violet's gone,
> The first-born child of the early sun.

With us, she is but a winter's flower;
The snow on the hills cannot blast her bower,
And she lifts up her dewy eye of blue
To the youngest sky of the selfsame hue.

And when the Spring comes, with her host
Of flowers, that flower beloved the most
Shrinks from the crowd that may confuse
Her heavenly odour, and virgin hues.

Pluck the others, but still remember
Their herald out of dim December,—
The morning star of all the flowers,
The pledge of daylight's lengthened hours,
Nor, midst the roses, e'er forget
The virgin, virgin violet." *

3. It is the queen, not only of the violet tribe, but of all low-growing flowers, in sweetness of scent—variously applicable and serviceable in domestic economy :—the. scent of the lily of the valley seems less capable of preservation or use.

But, respecting these perpetual beneficences and benignities of the sacred, as opposed to the malignant, herbs, whose poisonous power is for the most part re-

---

* A careless bit of Byron's, (the last song but one in the 'Deformed Transformed'); but Byron's most careless work is better, by its innate energy, than other people's most laboured. I suppress, in some doubts about my 'digamma,' notes on the Greek violet and the Ion of Euripides;—which the reader will perhaps be good enough to fancy a serious loss to him, and supply for himself.

strained in them, during their life, to their juices or dust, and not allowed sensibly to pollute the air, I should like the scholar to re-read pp. 251, 252 of vol. i., and then to consider with himself what a grotesquely warped and gnarled thing the modern scientific mind is, which fiercely busies itself in venomous chemistries that blast every leaf from the forests ten miles round; and yet cannot tell us, nor even think of telling us, nor does even one of its pupils think of asking it all the while, how a violet throws off her perfume!—far less, whether it might not be more wholesome to 'treat' the air which men are to breathe in masses, by administration of vale-lilies and violets, instead of charcoal and sulphur!

The closing sentence of the first volume just now referred to—p. 254—should also be re-read; it was the sum of a chapter I had in hand at that time on the Substances and Essences of Plants—which never got finished;—and in trying to put it into small space, it has become obscure: the terms "logically inexplicable" meaning that no words or process of comparison will define scents, nor do any traceable modes of sequence or relation connect them; each is an independent power, and gives a separate impression to the senses. Above all, there is no logic of pleasure, nor any assignable reason for the difference, between loathsome and delightful scent, which makes the fungus foul and the vervain sacred: but one practical conclusion I (who am in all final ways the most prosaic and practical of human

creatures) do very solemnly beg my readers to meditate;
namely, that although not recognized by actual offensive-
ness of scent, there is no space of neglected land which
is not in some way modifying the atmosphere of *all the
world*,—it may be, beneficently, as heath and pine,—it
may be, malignantly, as Pontine marsh or Brazilian
jungle; but, in one way or another, for good and evil
constantly, by day and night, the various powers of life
and death in the plants of the desert are poured into the
air, as vials of continual angels : and that no words, no
thoughts can measure, nor imagination follow, the possible change for good which energetic and tender care of
the wild herbs of the field and trees of the wood might
bring, in time, to the bodily pleasure and mental power
of Man.

32. II. Viola Psyche. Ophelia's Pansy.

The wild heart's-ease of Europe; its proper colour an
exquisitely clear purple in the upper petals, gradated
into deep blue in the lower ones; the centre, gold. Not
larger than a violet, but perfectly formed, and firmly set
in all its petals. Able to live in the driest ground;
beautiful in the coast sand-hills of Cumberland, following the wild geranium and burnet rose : and distinguished thus by its power of life, in waste and dry
places, from the violet, which needs kindly earth and
shelter.

Quite one of the most lovely things that Heaven has
made, and only degraded and distorted by any human

interference; the swollen varieties of it produced by
cultivation being all gross in outline and coarse in colour
by comparison.

It is badly drawn even in the ' Flora Danica,' No. 623,
considered there apparently as a species escaped from
gardens; the description of it being as follows:—

" Viola tricolor hortensis repens, flore purpureo et
cœruleo, C. B. P., 199." (I don't know what C. B. P.
means.) " Passim, juxta villas."

" Viola tricolor, caule triquetro diffuso, foliis oblongis
incisis, stipulis pinnatifidis," Linn. Systema Naturæ, 185.

33. "Near the country farms"—does the Danish
botanist mean?—the more luxuriant weedy character
probably acquired by it only in such neighbourhood;
and, I suppose, various confusion and degeneration possible to it beyond other plants when once it leaves its
wild home. It is given by Sibthorpe from the Trojan
Olympus, with an exquisitely delicate leaf; the flower described as " triste et pallide violaceus," but coloured in
his plate full purple; and as he does not say whether he
went up Olympus to gather it himself, or only saw it
brought down by the assistant whose lovely drawings
are yet at Oxford, I take leave to doubt his epithets.
That this should be the only Violet described in a ' Flora
Græca' extending to ten folio volumes, is a fact in
modern scientific history which I must leave the Professor of Botany and the Dean of Christ Church to
explain.

34. The English varieties seem often to be yellow in the lower petals, (see Sowerby's plate, 1287 of the old edition), crossed, I imagine, with Viola Aurea, (but see under Viola Rupestris, No. 12); the names, also, varying between tricolor and bicolor—with no note anywhere of the three colours, or two colours, intended!

The old English names are many.—' Love in idleness,' —making Lysander, as Titania, much wandering in mind, and for a time mere ' Kits run the street' (or run the wood?)—"Call me to you" (Gerarde, ch. 299, Sowerby, No. 178), with ' Herb Trinity,' from its three colours, blue, purple, and gold, variously blended in different countries? ' Three faces under a hood ' describes the English variety only. Said to be the ancestress of all the florists' pansies, but this I much doubt, the next following species being far nearer the forms most chiefly sought for.

35. III. Viola Alpina. ' Freneli's Pansy'—my own name for it, from Gotthelf's Freneli, in 'Ulric the Farmer'; the entirely pure and noble type of the Bernese maid, wife, and mother.

The pansy of the Wengern Alp in specialty, and of the higher, but still rich, Alpine pastures. Full dark-purple; at least an inch across the expanded petals; I believe, the ' Mater Violarum' of Gerarde; and true black violet of Virgil, remaining in Italian ' Viola Mammola ' (Gerarde, ch. 298).

36. IV. Viola Aurea. Golden Violet. Biflora usu-

ally; but its brilliant yellow is a much more definite characteristic; and needs insisting on, because there is a 'Viola lutea' which is not yellow at all; named so by the garden florists. My Viola aurea is the Rock-violet of the Alps; one of the bravest, brightest, and dearest of little flowers. The following notes upon it, with its summer companions, a little corrected from my diary of 1877, will enough characterize it.

"*June 7th.*—The cultivated meadows now grow only dandelions—in frightful quantity too; but, for wild ones, primula, bell gentian, golden pansy, and anemone,—Primula farinosa in mass, the pansy pointing and vivifying in a petulant sweet way, and the bell gentian here and there deepening all,—as if indeed the sound of a deep bell among lighter music.

"Counted in order, I find the effectively constant flowers are eight;* namely,

"1. The golden anemone, with richly cut large leaf; primrose colour, and in masses like primrose, studded through them with bell gentian, and dark purple orchis.

"2. The dark purple orchis, with bell gentian in equal quantity, say six of each in square yard, broken by sparklings of the white orchis and the white grass-flower; the richest piece of colour I ever saw, touched with gold by the geum.

---

* Nine; I see that I missed count of P. farinosa, the most abundant of all.

"3 and 4. These will be white orchis and the grass flower.*

"5. Geum—everywhere, in deep, but pure, gold, like pieces of Greek mosaic.

" 6. Soldanella, in the lower meadows, delicate, but not here in masses.

" 7. Primula Alpina, divine in the rock clefts, and on the ledges changing the grey to purple,—set in the dripping caves with

" 8. Viola (pertinax—pert); I want a Latin word for various studies—failures all—to express its saucy little stuck-up way, and exquisitely trim peltate leaf. I never saw such a lovely perspective line as the pure front leaf profile. Impossible also to get the least of the spirit of its lovely dark brown fibre markings. Intensely golden these dark fibres, just browning the petal a little between them."

And again in the defile of Gondo, I find " Viola (saxatilis ?) name yet wanted ;—in the most delicate studding of its round leaves, like a small fern more than violet, and bright sparkle of small flowers in the dark dripping hollows. Assuredly delights in shade and distilling moisture of rocks."

---

* " A feeble little quatrefoil—growing one on the stem, like a Parnassia. and looking like a Parnassia that had dropped a leaf. I think it drops one of its own four, mostly. and lives as three-fourths of itself, for most of its time. Stamens pale gold. Root-leaves, three or four, grass-like ; growing among the moist moss chiefly."

I found afterwards a much larger yellow pansy on the
Yorkshire high limestones ; with vigorously black crow-
foot marking on the lateral petals.

37. V. Viola Montana. Mountain Violet.

Flora Danica, 1329. Linnæus, No. 13, " Caulibus
erectis, foliis cordato-lanceolatis, floribus serioribus apeta-
lis," *i.e.*, on erect stems, with leaves long heart-shape,
and its later flowers without petals—not a word said of
its earlier flowers which have got those unimportant ap-
pendages! In the plate of the Flora it is a very perfect
transitional form between violet and pansy, with beautifully firm and well-curved leaves, but the colour of blossom very pale. " In subalpinis Norvegiæ passim," all
that we are told of it, means I suppose, in the lower
Alpine pastures of Norway; in the Flora Suecica, p.
306, habitat in Lapponica, juxta Alpes.

3S. VI. Viola Mirabilis. Flora Danica, 1045. A
small and exquisitely formed flower in the balanced
cinquefoil intermediate between violet and pansy, but
with large and superbly curved and pointed leaves. It
is a mountain violet, but belonging rather to the mountain woods than meadows. "In sylvaticis in Toten,
Norvegiæ."

Loudon, 3056, " Broad-leaved : Germany."

Linnæus, Flora Suecica, 789, says that the flowers of
it which have perfect corolla and full scent often bear
no seed, but that the later ' cauline' blossoms, without
petals, are fertile. " Caulini vero apetali fertiles sunt,
et seriores. Habitat passim Upsaliæ."

I find this, and a plurality of other species, indicated by Linnæus as having triangular stalks, "caule triquetro," meaning, I suppose, the kind sketched in Figure 1 above.

39. VII. VIOLA ARVENSIS. Field Violet. Flora Danica, 1748. A coarse running weed; nearly like Viola Cornuta, but feebly lilac and yellow in colour. In dry fields, and with corn.

Flora Suecica, 791; under titles of Viola 'tricolor' and 'bicolor arvensis,' and Herba Trinitatis. Habitat ubique in *sterilibus* arvis: "Planta vix datur in qua evidentius perspicitur generationis opus, quam in hujus cavo apertoque stigmate."

It is quite undeterminable, among present botanical instructors, how far this plant is only a rampant and over-indulged condition of the true pansy (Viola Psyche); but my own scholars are to remember that the true pansy is full purple and blue with golden centre; and that the disorderly field varieties of it, if indeed not scientifically distinguishable, are entirely separate from the wild flower by their scattered form and faded or altered colour. I follow the Flora Danica in giving them as a distinct species.

40. VIII. VIOLA PALUSTRIS. Marsh Violet. Flora Danica, 83. As there drawn, the most finished and delicate in form of all the violet tribe; warm white, streaked with red; and as pure in outline as an oxalis, both in flower and leaf: it is like a violet imitating oxalis and anagallis.

3

In the Flora Suecica, the petal-markings are said to be black; in ' Viola lactea' a connected species, (Sowerby, 45,) purple. Sowerby's plate of it under the name ' palustris' is pale purple veined with darker; and the spur is said to be 'honey-bearing,' which is the first mention I find of honey in the violet. The habitat given, sandy and turfy heaths. It is said to grow plentifully near Croydon.

Probably, therefore, a violet belonging to the chalk, on which nearly all herbs that grow wild—from the grass to the bluebell—are singularly sweet and pure. I hope some of my botanical scholars will take up this question of the effect of different rocks on vegetation, not so much in bearing different species of plants, as different characters of each species.*

41. IX. VIOLA SECLUSA. Monk's Violet. "Hirta," Flora Danica, 618, " In fruticetis raro." A true wood violet, full but dim in purple. Sowerby, 894, makes it paler. The leaves very pure and severe in the Danish one;—longer in the English. " Clothed on both sides with short, dense, hoary hairs."

Also belongs to chalk or limestone only (Sowerby).

X. VIOLA CANINA. Dog Violet. I have taken it for analysis in my two plates, because its grace of form is too much despised, and we owe much more of the beauty

---

* The great work of Lecoq, ' Geographie Botanique,' is of priceless value; but treats all on too vast a scale for our purposes.

of spring to it, in English mountain ground, than to the
. Regina.

XI. VIOLA CORNUTA. Cow Violet. Enough described
already.

XII. VIOLA RUPESTRIS. Crag Violet. On the high
limestone moors of Yorkshire, perhaps only an English
form of Viola Aurea, but so much larger, and so different in habit—growing on dry breezy downs, instead of
in dripping caves—that I allow it, for the present, separate name and number.*

42. 'For the present,' I say all this work in ' Proserpina' being merely tentative, much to be modified by
future students, and therefore quite different from that
of ' Deucalion,' which is authoritative as far as it reaches,
and will stand out like a quartz dyke, as the sandy specu-
lations of modern gossiping geologists get washed away.

But in the meantime, I must again solemnly warn my
girl-readers against all study of floral genesis and diges-
tion. How far flowers invite, or require, flies to inter-
fere in their family affairs—which of them are carnivo-
rous—and what forms of pestilence or infection are most
favourable to some vegetable and animal growths,—let
them leave the people to settle who like, as Toinette says

* It is, I believe, Sowerby's Viola Lutea, 721 of the old edition, there
painted with purple upper petals ; but he says in the text, "Petals
either all yellow, or the two uppermost are of a blue purple, the rest
yellow with a blue tinge : very often the whole are purple."

of the Doctor in the ' Malade Imaginaire '—" y mettre le nez." I observe a paper in the last 'Contemporary Review,' announcing for a discovery patent to all mankind that the colours of flowers were made "to attract insects"! * They will next hear that the rose was made for the canker, and the body of man for the worm.

43. What the colours of flowers, or of birds, or of precious stones, or of the sea and air, and the blue mountains, and the evening and the morning, and the clouds of Heaven, were given for—they only know who can see them and can feel, and who pray that the sight and the love of them may be prolonged, where cheeks will not fade, nor sunsets die.

44. And now, to close, let me give you some fuller account of the reasons for the naming of the order to which the violet belongs, ' Cytherides.'

You see that the Uranides, are, as far as I could so gather them, of the pure blue of the sky; but the Cytherides of altered blue;—the first, Viola, typically purple; the second, Veronica, pale blue with a peculiar light; the third, Giulietta, deep blue, passing strangely into a subdued green before and after the full life of the flower.

All these three flowers have great strangenesses in them, and weaknesses; the Veronica most wonderful in

---

* Did the wretch never hear bees in a lime tree then, or ever see one on a star gentian?

its connection with the poisonous tribe of the foxgloves; the Giulietta, alone among flowers in the action of the shielding leaves; and the Viola, grotesque and inexplicable in its hidden structure, but the most sacred of all flowers to earthly and daily Love, both in its scent and glow.

Now, therefore, let us look completely for the meaning of the two leading lines,—

> " Sweeter than the lids of Juno's eyes,
> Or Cytherea's breath."

45. Since, in my present writings, I hope to bring into one focus the pieces of study fragmentarily given during past life, I may refer my readers to the first chapter of the ' Queen of the Air' for the explanation of the way in which all great myths are founded, partly on physical, partly on moral fact,—so that it is not possible for persons who neither know the aspect of nature, nor the constitution of the human soul, to understand a word of them. Naming the Greek gods, therefore, you have first to think of the physical power they represent. When Horace calls Vulcan ' Avidus,' he thinks of him as the power of Fire; when he speaks of Jupiter's red right hand, he thinks of him as the power of rain with lightning; and when Homer speaks of Juno's dark eyes, you have to remember that she is the softer form of the rain power, and to think of the fringes of the rain-cloud across the light of the horizon. Gradually the idea becomes per-

sonal and human in the "Dove's eyes within thy locks," *
and "Dove's eyes by the river of waters" of the Song of
Solomon.

46. "Or Cytherea's breath,"—the two thoughts of softest glance, and softest kiss, being thus together associated
with the flower: but note especially that the Island of
Cythera was dedicated to Venus because it was the chief,
if not the only Greek island, in which the purple fishery
of Tyre was established; and in our own minds should
be marked not only as the most southern fragment of
true Greece, but the virtual continuation of the chain of
mountains which separate the Spartan from the Argive
territories, and are the natural home of the brightest
Spartan and Argive beauty which is symbolized in Helen.

47. And, lastly, in accepting for the order this name
of Cytherides, you are to remember the names of Viola
and Giulietta, its two limiting families, as those of Shakspeare's two most loving maids—the two who love simply, and to the death: as distinguished from the greater
natures in whom earthly Love has its due part, and no
more; and farther still from the greatest, in whom the
earthly love is quiescent, or subdued, beneath the thoughts
of duty and immortality.

It may be well quickly to mark for you the levels of

---

* Septuagint, "the eyes of doves out of thy silence." Vulgate,
"the eyes of doves, besides that which is hidden in them." Meaning
—the *dim* look of love, beyond all others in sweetness.

loving temper in Shakspeare's maids and wives, from the greatest to the least.

48. 1. Isabel. All earthly love, and the possibilities of it, held in absolute subjection to the laws of God, and the judgments of His will. She is Shakspeare's only 'Saint.' Queen Catherine, whom you might next think of, is only an ordinary woman of trained religious temper:—her maid of honour gives Wolsey a more Christian epitaph.

2. Cordelia. The earthly love consisting in diffused compassion of the universal spirit; not in any conquering, personally fixed, feeling.

> ' Mine enemy's dog,
> Though he had bit me, should have stood that night
> Against my fire."

These lines are spoken in her hour of openest direct expression ; and are *all* Cordelia.

Shakspeare clearly does not mean her to have been supremely beautiful in person; it is only her true lover who calls her 'fair' and 'fairest'—and even that, I believe, partly in courtesy, after having the instant before offered her to his subordinate duke; and it is only *his* scorn of her which makes France fully care for her.

> " Gods, Gods, 'tis strange that from their cold neglect
> My love should kindle to inflamed respect !"

Had she been entirely beautiful, he would have honoured her as a lover should, even before he saw her despised;

nor would she ever have been so despised—or by her
father, misunderstood.   Shakspeare himself does not pretend to know where her girl-heart was,—but I should
like to hear how a great actress would say the " Peace be
with Burgundy !"

3. Portia. The maidenly passion now becoming great,
and chiefly divine in its humility, is still held absolutely
subordinate to duty; no thought of disobedience to her
dead father's intention is entertained for an instant, though
the temptation is marked as passing, for that instant, before her crystal strength.   Instantly, in her own peace,
she thinks chiefly of her lover's ;—she is a perfect Christian wife in a moment, coming to her husband with the
gift of perfect Peace,—

> " Never shall you lie by Portia's side
> With an unquiet soul."

She is highest in intellect of all Shakspeare's women,
and this is the root of her modesty ; her ' unlettered girl'
is like Newton's simile of the child on the sea-shore.
Her perfect wit and stern judgment are never disturbed
for an instant by her happiness : and the final key to her
character is given in her silent and slow return from
Venice, where she stops at every wayside shrine to pray.

4. Hermione.   Fortitude and Justice personified, with
unwearying affection.   She is Penelope, tried by her husband's fault as well as error.

5. Virgilia.   Perfect type of wife and mother, but

without definiteness of character, nor quite strength of
intellect enough entirely to hold her husband's heart.
Else, she had saved him: he would have left Rome in his
wrath—but not her. Therefore, it is his mother only
who bends him: but she cannot save.

6. Imogen. The ideal of grace and gentleness; but
weak; enduring too mildly, and forgiving too easily.
But the piece is rather a pantomime than play, and it is
impossible to judge of the feelings of St. Columba, when
she must leave the stage in half a minute after mistaking
the headless clown for headless Arlecchino.

7. Desdemona, Ophelia, Rosalind. They are under
different conditions from all the rest, in having entirely
heroic and faultless persons to love. I can't class them,
therefore,—fate is too strong, and leaves them no free
will.

8. Perdita, Miranda. Rather mythic visions of maiden
beauty than mere girls.

9. Viola and Juliet. Love the ruling power in the entire character: wholly virginal and pure, but quite earthly, and recognizing no other life than his own. Viola is,
however, far the noblest. Juliet will die unless Romeo
loves *her:* "If he be wed, the grave is like to be my wed-
ding bed;" but Viola is ready to die for the happiness
of the man who does *not* love her; faithfully doing his
messages to her rival, whom she examines strictly for his
sake. It is not in envy that she says, "Excellently done,
—if God did all." The key to her character is given in

the least selfish of all lover's songs, the one to which the
Duke bids her listen :

> " Mark it, Cesario,—it is old and plain,
>   The spinsters and the knitters in the sun,
>   And the free maids, that *weave their thread with bones,*
>   Do use to chaunt it."

(They, the unconscious Fates, weaving the fair vanity of
life with death); and the burden of it is—

> " My part of Death, no one so true
>   Did share it."

Therefore she says, in the great first scene, " Was not
this love indeed?" and in the less heeded closing one,
her heart then happy with the knitters in the *sun,*

> " And all those sayings will I over-swear,
>   And all those swearings keep as true in soul
>   As doth that orbed continent the Fire
>   That severs day from night."

Or, at least, did once sever day from night,—and perhaps
does still in Illyria.   Old England must seek new images
for her loves from gas and electric sparks,—not to say
furnace fire.

I am obliged, by press of other work, to set down these
notes in cruel shortness : and many a reader may be disposed to question utterly the standard by which the
measurement is made.   It will not be found, on reference

to my other books, that they encourage young ladies to
go into convents; or undervalue the dignity of wives and
mothers. But, as surely as the sun *does* sever day from
night, it will be found always that the noblest and loveliest women are dutiful and religious by continual nature;
and their passions are trained to obey them; like their
dogs. Homer, indeed, loves Helen with all his heart,
and restores her, after all her naughtiness, to the queen-
ship of her household; but he never thinks of her as
Penelope's equal, or Iphigenia's. Practically, in daily
life, one often sees married women as good as saints; but
rarely, I think, unless they have a good deal to bear from
their husbands. Sometimes also, no doubt, the husbands
have some trouble in managing St. Cecilia or St. Eliza-
beth; of which questions I shall be obliged to speak
more seriously in another place: content, at present, if
English maids know better, by Proserpina's help, what
Shakspeare meant by the dim, and Milton by the glowing,
violet.

CHAPTER II.

. (Written in early June, 1881.)

1. On the rocks of my little stream, where it runs, or leaps, through the moorland, the common Pinguicula is now in its perfectest beauty; and it is one of the off-shoots of the violet tribe which I have to place in the minor collateral groups of Viola very soon, and must not put off looking at it till next year.

There are three varieties given in Sowerby: 1. Vulgaris, 2. Greater-flowered, and 3. Lusitanica, white, for the most part, pink, or 'carnea,' sometimes: but the proper colour of the family is violet, and the perfect form of the plant is the 'vulgar' one. The larger-flowered variety is feebler in colour, and ruder in form: the white Spanish one, however, is very lovely, as far as I can judge from Sowerby's (*old* Sowerby's) pretty drawing.

The 'frequent' one (I shall usually thus translate 'vulgaris'), is not by any means so 'frequent' as the Queen violet, being a true wild-country, and mostly Alpine, plant; and there is also a real 'Pinguicula Alpina,'

which we have not in England, who might be the Regina, if the group were large enough to be reigned over:
but it is better not to affect Royalty among these confused, intermediate, or dependent families.

2. In all the varieties of Pinguicula, each blossom has
one stalk only, growing from the *ground ;* and you may
pull all the leaves away from the base of it, and keep
the flower only, with its bunch of short fibrous roots,
half an inch long; looking as if bitten at the ends.
Two flowers, characteristically,—three and four very
often,—spring from the same root, in places where it
grows luxuriantly; and luxuriant growth means that
clusters of some twenty or thirty stars may be seen on
the surface of a square yard of boggy ground, quite to
its mind; but its real glory is in harder life, in the cran-
nies of well-wetted rock.

3. What I have called 'stars' are irregular clusters of
approximately, or tentatively, five alocine ground leaves,
of very pale green,—they may be six or seven, or more,
but always run into a rudely pentagonal arrangement,
essentially first trine, with two succeeding above.
Taken as a whole the *plant* is really a main link between
violets and Droseras; but the *flower* has much more
violet than Drosera in the make of it,—spurred, and *five-
petaled,** and held down by the top of its bending stalk

---

* When I have the chance, and the time, to submit the proofs of
'Proserpina' to friends who know more of Botany than I, or have

as a violet is; only its upper two petals are not reverted
—the calyx, of a dark soppy green, holding them down,
with its three front sepals set exactly like a strong tri-
dent, its two backward sepals clasping the spur.   There

---

kindness enough to ascertain debateable things for me, I mean in
future to do so,—using the letter A to signify Amicus, generally;
with acknowledgment by name, when it is permitted, of especial
help or correction.   Note first of this kind: I find here on this word,
'five-petaled,' as applied to Pinguicula, "Qy. two-lipped? it is monopetalous, and monosepalous, the calyx and corolla being each all in
one piece."
    Yes; and I am glad to have the observation inserted.   But my
term, 'five petaled,' must stand.   For the question with me is always
first, not how the petals are connected, but how many they are.
Also I have accepted the term petal—but never the word lip—as applied to flowers.   The generic term 'Labiatæ' is cancelled in 'Proserpina,' 'Vestales' being substituted; and these flowers, when I come
to examine them, are to be described, not as divided into two lips,
but into hood, apron, and side-pockets.   Farther, the depth to which
either calyx or corolla is divided, and the firmness with which the
petals are attached to the torus, may, indeed, often be an important
part of the plant's description, but ought not to be elements in its
definition.   Three petaled and three-sepaled, four-petaled and four-
sepaled, five-petaled and five-sepaled, etc., etc., are essential—with
me, primal—elements of definition; next, whether resolute or stellar
in their connection; next, whether round or pointed, etc.   Fancy,
for instance, the fatality to a rose of pointing its petals, and to a lily,
of rounding them!   But how deep cut, or how hard holding, is
quite a minor question.
    Farther, that all plants *are* petaled and sepaled, and never mere
cups in saucers, is a great fact, not to be dwelt on in a note.

are often six sepals, four to the front, but the normal
number is five. Tearing away the calyx, I find the
flower to have been held by it as a lion might hold his
prey by the loins if he missed its throat; the blue petals
being really campanulate, and the flower best described
as a dark bluebell, seized and crushed almost flat by its
own calyx in a rage. Pulling away now also the upper
petals, I find that what are in the violet the lateral and
well-ordered fringes, are here thrown mainly on the
lower (largest) petal near its origin, and opposite the
point of the seizure by the calyx, spreading from this
centre over the surface of the lower petals, partly like an
irregular shower of fine Venetian glass broken, partly
like the wild-flung Medusa-like embroidery of the white
Lucia.*

4. The calyx is of a dark *soppy* green, I said; like
that of sugary preserved citron; the root leaves are of
green just as soppy, but pale and yellowish, as if they
were half decayed; the edges curled up and, as it were,
water-shrivelled, as one's fingers shrivel if kept too long
in water. And the whole plant looks as if it had been a
violet unjustly banished to a bog, and obliged to live
there—not for its own sins, but for some Emperor
Pansy's, far away in the garden,—in a partly boggish,

---

* Our 'Lucia Nivea,' 'Blanche Lucy;' in present botany, Bog
bean! having no connection whatever with any manner of bean, but
only a slight resemblance to bean-*leaves* in its own lower ones. Compare Ch. IV. § 11.

partly hoggish manner, drenched and desolate; and with something of demoniac temper got into its calyx, so that it quarrels with, and bites the corolla;—something of gluttonous and greasy habit got into its leaves; a discomfortable sensuality, even in its desolation. Perhaps a penguin-ish life would be truer of it than a piggish, the *nest* of it being indeed on the rock, or morassy rock-investiture, like a sea-bird's on her rock ledge.

5. I have hunted through seven treatises on Botany, namely, Loudon's Encyclopædia, Balfour, Grindon, Oliver, Baxter of Oxford, Lindley ('Ladies' Botany'), and Figuier, without being able to find the meaning of 'Lentibulariaceæ,' to which tribe the Pinguicula is said by them all (except Figuier) to belong. It may perhaps be in Sowerby :* but these above-named treatises are precisely of the kind with which the ordinary scholar must be content: and in all of them he has to learn this long, worse than useless, word, under which he is betrayed into classing together two orders naturally quite distinct, the Butterworts and the Bladderworts.

Whatever the name may mean—it is bad Latin. There is such a word as Lenticularis—there is no Lentibularis; and it must positively trouble us no longer.†

---

* It is not. (Resolute negative from A., unsparing of time for me; and what a state of things it all signifies !)

† With the following three notes, 'A' must become a definitely and gratefully interpreted letter. I am indebted for the first, conclusive

The Butterworts are a perfectly distinct group—whether small or large, always recognizable at a glance. Their proper Latin name will be Pinguicula, (plural Pinguiculæ,)—their English, Bog-Violet, or, more familiarly, Butterwort; and their French, as at present, *Grassette.*

---

in itself, but variously supported and confirmed by the two following, to R. J. Mann, Esq., M.D., long ago a pupil of Dr. Lindley's, and now on the council of Whitelands College, Chelsea:—for the second, to Mr. Thomas Moore, F.L.S., the kind Keeper of the Botanic Garden at Chelsea; for the third, which will be farther on useful to us, to Miss Kemm, the botanical lecturer at Whitelands.

(1) There is no explanation of Lentibulariaceæ in Lindley's ' Vegetable Kingdom.' He was not great in that line. The term is, however, taken from *Lenticula,* the lentil, in allusion to the lentil-shaped air-bladders of the typical genus *Utricularia.*

The change of the c into b may possibly have been made only from some euphonic fancy of the contriver of the name, who, I think, was Rich.

But I somewhat incline myself to think that the *tibia,* a pipe or flute, may have had something to do with it. The *tibia* may possibly have been diminished into a little pipe by a stretch of licence, and have become *tibula:* [but *tibulus* is a kind of pine tree in Pliny]; when *Len tibula* would be the lens or lentil-shaped pipe or bladder. I give you this only for what it is worth. The *lenticula,* as a derivation, is reliable and has authority.

Lenticula, a lentil, a freckly eruption; *lenticularis,* lentil-shaped; so the nat. ord. ought to be (if this be right) *lenticulariaceœ.*

(2)   BOTANIC GARDENS, CHELSEA, *Feb.* 14, 1882.

Lentibularia is an old generic name of Tournefort's, which has been superseded by *utricularia,* but, oddly enough, has been retained

4

The families to be remembered will be only five, namely,

1. Pinguicula Major, the largest of the group. As bog plants, Ireland may rightly claim the noblest of them, which certainly grow there luxuriantly, and not (I believe) with us. Their colour is, however, more broken and less characteristic than that of the following species.

2. Pinguicula Violacea: Violet-coloured Butterwort, (instead of 'vulgaris,') the common English and Swiss kind above noticed.

3. Pinguicula Alpina: Alpine Butterwort, white and much smaller than either of the first two families; the

---

in the name of the order *lentibulareæ;* but it probably comes from *lenticula,* which signifies the little root bladders, somewhat resembling lentils.

(3) 'Manual of Scientific Terms,' Stormonth, p. 234.

Lentibulariaceæ, neuter, plural.

(*Lenticula,* the shape of a lentil; from *lens,* a lentil.) The Butterwort family, an order of plants so named from the lenticular shape of the air-bladders on the branches of utricularia, one of the genera. (But observe that the *Butterworts* have nothing of the sort, any of them.—R.)

Loudon.—"Floaters."

Lindley.—"Sometimes with whorled vesicles."

In Nuttall's Standard (?) Pronouncing Dictionary, it is given,—

Lenticulareæ, a nat. ord. of marsh plants, which thrive in water or marshes.

spur especially small, according to D. 453. Much rarer, as well as smaller, than the other varieties in Southern Europe. "In Britain, known only upon the moors of Rosehaugh, Ross-shire, where the progress of cultivation seems likely soon to efface it." (Grindon.)

4. Pinguicula Pallida: Pale Butterwort. From Sowerby's drawing, (135, vol. iii.,) it would appear to be the most delicate and lovely of all the group. The leaves, " like those of other species, but rather more delicate and pellucid, reticulated with red veins, and much involute in the margin. Tube of the corolla, yellow, streaked with red, (the streaks like those of a pansy); the petals, pale violet. It much resembles Villosa, (our Minima, No. 5,) in many particulars, the stem being hairy, and in the lower part the hairs tipped with a viscid fluid, like a sundew. But the Villosa has a slender sharp spur; and in this the spur is blunt and thick at the end." (Since the hairy stem is not peculiar to Villosa, I take for her, instead, the epithet Minima, which is really definitive.)

The pale one is commonly called ' Lusitanica,' but I find no direct notice of its Portuguese habitation. Sowerby's plant came from Blandford, Dorsetshire ; and Grindon says it is frequent in Ireland, abundant in Arran, and extends on the western side of the British island from Cornwall to Cape Wrath. My epithet, Pallida, is secure, and simple, wherever the plant is found.

5. Pinguicula Minima: Least Butterwort; in D. 1021 called Villosa, the *scape* of it being hairy. I have not yet got rid of this absurd word 'scape,' meaning, in botanist's Latin, the flower-stalk of a flower growing out of a cluster of leaves on the ground. It is a bad corruption of 'sceptre,' and especially false and absurd, because a true sceptre is necessarily branched.* In 'Proserpina,' when it is spoken of distinctively, it is called 'virgula' (see vol. i., pp. 146, 147, 151, 152). The hairs on the virgula are in this instance so minute, that even with a lens I cannot see them in the Danish plate: of which Fig. 3 is a rough translation into woodcut, to show the grace and

FIG. III. mien of the little thing. The trine leaf cluster is characteristic, and the folding up of the leaf edges. The flower, in the Danish plate, full purple. Abundant in east of *Finmark* (Finland?), but *always growing in marsh moss,* (Sphagnum palustre).

6. I call it 'Minima' only, as the least of the five here named; without putting forward any claim for it to be the smallest pinguicula that ever was or will be. In such sense only, the epithets minima or maxima are to be understood when used in 'Proserpina': and so also,

---

* More accurately, shows the pruned roots of branches,—ἐπειδὴ πρῶτα τομὴν ἐν ὄρεσσι λέλοιπεν. The *pruning* is the mythic expression of the subduing of passion by rectorial law.

every statement and every principle is only to be understood as true or tenable, respecting the plants which the writer has seen, and which he is sure that the reader can easily see : liable to modification to any extent by wider experience ; but better first learned securely within a narrow fence, and afterwards trained or fructified, along more complex trellises.

7. And indeed my readers—at least, my newly found readers—must note always that the only power which I claim for any of my books, is that of being right and true as far as they reach. None of them pretend to be Kosmoses ;—none to be systems of Positivism or Negativism, on which the earth is in future to swing instead of on its old worn-out poles ;—none of them to be works of genius ;—none of them to be, more than all true work *must* be, pious ;—and none to be, beyond the power of common people's eyes,* ears, and noses, 'æsthetic.' They tell you that the world is *so* big, and can't be made bigger—that you yourself are also so big, and can't be made bigger, however you puff or bloat yourself; but that, on modern mental nourishment, you may very easily be made smaller. They tell you that two and two are four, that ginger is hot in the mouth, that roses are red, and smuts black. Not themselves assuming to be

---

* The bitter sorrow with which I first recognized the extreme rarity of finely-developed organic sight is expressed enough in the lecture on the Mystery of Life, added in the large edition of 'Sesame and Lilies.'

pious, they yet assure you that there is such a thing as
piety in the world, and that it is wiser than impiety; and
not themselves pretending to be works of genius, they
yet assure you that there is such a thing as genius in the
world, and that it is meant for the light and delight of
the world.

8. Into these repetitions of remarks on my work,
often made before, I have been led by an unlucky author
who has just sent me his book, advising me that it is
"neither critical nor sentimental" (he had better have
said in plain English "without either judgment or feeling"), and in which nearly the first sentence I read is—
"Solomon with all his acuteness was not wise enough to
. . . etc., etc., etc." ('give the Jews the British constitution,' I believe the man means.)   He is not a whit
more conceited than Mr. Herbert Spencer, or Mr. Goldwin Smith, or Professor Tyndall,—or any lively London
apprentice out on a Sunday; but this general supercili-
ousness with respect to Solomon, his Proverbs, and his
politics, characteristic of the modern Cockney, Yankee,
and Anglicised Scot, is a difficult thing to deal with for
us of the old school, who were well whipped when we
were young; and have been in the habit of occasionally
ascertaining our own levels as we grew older, and of
recognizing that, here and there, somebody stood higher,
and struck harder.

9. A difficult thing to deal with, I feel more and
more, hourly, even to the point of almost ceasing to

write; not only every feeling I have, but, of late, even
every word I use, being alike inconceivable to the inso-
lence. and unintelligible amidst the slang, of the modern
London writers.   Only in the last magazine I took up, I
found an article by Mr. Goldwin Smith on the Jews (of
which the gist—as far as it had any—was that we had
better give up reading the Bible), and in the text of
which I found the word 'tribal' repeated about ten times
in every page.   Now, if 'tribe' makes 'tribal,' tube
must make tubal, cube, cubal, and gibe, gibal; and I
suppose we shall next hear of tubal music, cubal min-
erals, and gibal conversation!   And observe how all this
bad English leads instantly to blunder in thought, prolonged indefinitely.   The Jewish Tribes are not separate
races, but the descendants of brothers.   The Roman
Tribes, political divisions; essentially Trine : and the
whole force of the word Tribune vanishes, as soon as the
ear is wrung into acceptance of his lazy innovation by
the modern writer.   Similarly, in the last elements of
mineralogy I took up. the first order of crystals was
called 'tesseral'; the writer being much too fine to call
them 'four-al,' and too much bent on distinguishing
himself from all previous writers to call them cubic.

10. What simple schoolchildren, and sensible school-
masters, are to do in this atmosphere of Egyptian marsh,
which rains fools upon them like frogs, I can no more
with any hope or patience conceive;—but this finally I
repeat, concerning my own books, that they are written

in honest English, of good Johnsonian lineage, touched
here and there with colour of a little finer or Elizabethan
quality: and that the things they tell you are comprehensible by any moderately industrious and intelligent
person; and *accurate*, to a degree which the accepted
methods of modern science cannot, in my own particular
fields, approach.

11. Of which accuracy, the reader may observe for
immediate instance, my extrication for him, from among
the uvularias, of these five species of the Butterwort;
which, being all that need be distinctly named and remembered, *do* need to be first carefully distinguished,
and then remembered in their companionship. So alike
are they, that Gerarde makes no distinction among them;
but masses them under the general type of the frequent
English one, described as the second kind of his promis-
cuous group of 'Sanicle,' "which Clusius calleth Pinguicula; not before his time remembered, hath sundry small
thick leaves, fat and full of juice, being broad towards
the root and sharp towards the point, of a faint green
colour, and bitter in taste; out of the middest whereof
sprouteth or shooteth up a naked slender stalke nine
inches long, every stalke bearing one flower and no more,
sometimes white, and sometimes of a bluish purple colour,
fashioned like unto the common Monkshoods" (he means
Larkspurs) "called Consolida Regalis, having the like spur
or Lark's heel attached thereto." Then after describing a third kind of Sanicle—(Cortusa Mathioli, a large-

leaved Alpine Primula,) he goes on : " These plants are strangers in England ; their natural country is the alpish mountains of Helvetia. They grow in my garden, where they flourish exceedingly, except Butterwoort, which groweth in our English *squally* wet grounds,"—(' Squally,' I believe, here, from squalidus, though Johnson does not give this sense ; but one of his quotations from Ben Jonson touches it nearly: " Take heed that their new flowers and sweetness do not as much corrupt as the others' dryness and squalor,"—and note farther that the word ' squal.' in the sense of gust, is not pure English, but the Arabic ' Chuaul ' with an s prefixed :—the English word, a form of ' squeal,' meaning a child's cry, from Gothic ' Squæla ' and Icelandic ' squilla,' would scarcely have been made an adjective by Gerarde),—" and will not yield to any culturing or transplanting : it groweth especially in a field called Cragge Close, and at Crosbie Ravenswaithe, in Westmerland ; (West-*mere*-land you observe, not *mor*) upon Ingleborough Fells, twelve miles from Lancaster, and by Harwoode in the same county near to Blackburn : ten miles from Preston, in Anderness, upon the bogs and marish ground, and in the boggie meadows about Bishop's-Hatfield, and also in the fens in the way to Wittles Meare" (Roger Wildrake's Squattlesea Mere ?) " from Fendon, in Huntingdonshire." Where doubtless Cromwell ploughed it up, in his young days, pitilessly ; and in nowise pausing, as Burns beside his fallen daisy.

12. Finally, however, I believe we may accept its English name of 'Butterwort' as true Yorkshire, the more enigmatic form of 'Pigwilly' preserving the tradition of the flowers once abounding, with softened Latin name, in Pigwilly bottom, close to Force bridge, by Kendal. Gerarde draws the English variety as "Pingnicula sive Sanicula Eboracensis,—Butterwoort, or Yorkshire Sanicle;" and he adds: "The husbandmen's wives of Yorkshire do use to anoint the dugs of their kine with the fat and oilous juice of the herb Butterwort when they be bitten of any venomous worm, or chapped, rifted and hurt by any other means."

13. In Lapland it is put to much more certain use; "it is called Tätgrass, and the leaves are used by the inhabitants to make their 'tät miolk,' a preparation of milk in common use among them. Some fresh leaves are laid upon a filter, and milk, yet warm from the reindeer, is poured over them. After passing quickly through the filter, this is allowed to rest for one or two days until it becomes ascescent,* when it is found not to have separated from the whey, and yet to have attained much greater tenacity and consistence than it would have done otherwise. The Laplanders and Swedes are said to be extremely fond of this milk, which when once made, it is not necessary to renew the use of. the leaves, for we are told that a spoonful of it will turn another quantity of

---

* Lat. acesco, to turn sour.

warm milk, and make it like the first."* (Baxter, vol,
iii., No. 209.)

14. In the same page, I find quoted Dr. Johnston's
observation that "when specimens of this plant were
somewhat rudely pulled up, the flower-stalk, previously
erect, almost immediately began to bend itself backwards,
and formed a more or less perfect segment of a circle;
and so also, if a specimen is placed in the Botanic box,
you will in a short time find that the leaves have curled
themselves backwards, and now conceal the root by their
revolution."

I have no doubt that this elastic and wiry action is
partly connected with the plant's more or less predatory
or fly-trap character, in which these curiously degraded
plants are associated with Drosera. I separate them
therefore entirely from the Bladderworts, and hold them
to be a link between the Violets and the Droseraceæ,

---

* Withering quotes this as from Linnæus, and adds on authority
of a Mr. Hawkes, "This did not succeed when tried with cows' milk."
He also gives as another name, Yorkshire Sanicle; and says it is
called *earning grass* in Scotland. Linnæus says the juice will curdle
reindeer's milk. The name for rennet is *earning*, in Lincolnshire.
Withering also gives this note : "*Pinguis*, fat, from its effect in con-
GEALING milk."—(A.) Withering of course wrong : the name comes,
be the reader finally assured, from the fatness of the green leaf, quite
peculiar among wild plants, and fastened down for us in the French
word 'Grassette.' I have found the flowers also difficult to dry, in the
benighted early times when I used to think a dried plant useful ! See
closing paragraphs of the 4th chapter.—R.

placing them, however, with the Cytherides, as a subfamily, for their beautiful colour, and because they are indeed a grace and delight in ground which, but for them, would be painfully and rudely desolate.

CHAPTER III.

1. "THE Corolla of the Foxglove," says Dr. Lindley, beginning his account of the tribe at page 195 of the first volume of his 'Ladies' Botany,' "is a large inflated body (!), with its throat spotted with rich purple, and its border divided obliquely into five very short lobes, of which the two upper are the smaller; its four stamens are of unequal length, and its style is divided into two lobes at the upper end. A number of long hairs cover the ovary, which contains two cells and a great quantity of ovules.

"This" (sc. information) "will show you what is the usual character of the Foxglove tribe; and you will find that all the other genera referred to it in books agree with it essentially, although they differ in subordinate points. It is chiefly (A) in the form of the corolla, (B) in the number of the stamens, (C) in the consistence of the rind of the fruit, (D) in its form, (E) in the number of the seeds it contains, and (F) in the manner in which the sepals are combined, that these differences consist."

2. The enumerative letters are of my insertion—otherwise the above sentence is, word for word, Dr. Lindley's, —and it seems to me an interesting and memorable one

in the history of modern Botanical science. For it appears from the tenor of it, that in a scientific botanist's mind, six particulars, at least, in the character of a plant, are merely 'subordinate points,'—namely,

1. (F) The combination of its calyx,
2. (A) The shape of its corolla,
3. (B) The number of its stamens,
4. (D) The form of its fruit,
5. (C) The consistence of its shell,—and
6. (E) The number of seeds in it.

Abstracting, then, from the primary description, all the six inessential points, I find the three essential ones left are, that the style is divided into two lobes at the upper end, that a number of glandular hairs cover the ovary, and that this latter contains two cells.

3. None of which particulars concern any reasonable mortal, looking at a Foxglove, in the smallest degree. Whether hairs which he can't see are glandular or bristly, —whether the green knobs, which are left when the purple bells are gone, are divided into two lobes or two hundred,—and whether the style is split, like a snake's tongue, into two lobes, or like a rogue's, into any number—are merely matters of vulgar curiosity, which he needs a microscope to discover, and will lose a day of his life in discovering. But if any pretty young Proserpina, escaped from the Plutonic durance of London, and carried by the tubular process, which replaces Charon's boat, over

the Lune at Lancaster, cares to come and walk on the
Coniston hills in a summer morning, when the eyebright
is out on the high fields, she may gather, with a little
help from Brantwood garden, a bouquet of the entire
Foxglove tribe in flower, as it is at present defined, and
may see what they are like, altogether.

4. She shall gather : first, the Euphrasy, which makes
the turf on the brow of the hill glitter as if with new-
fallen manna ; then, from one of the blue clusters on the
top of the garden wall, the common bright blue Speedwell; and, from the garden bed beneath, a dark blue
spire of Veronica spicata ; then, at the nearest opening
into the wood, a little foxglove in its first delight of
shaking out its bells ; then—what next does the Doctor
say ?—a snapdragon ? we must go back into the garden
for that—here is a goodly crimson one, but what the
little speedwell will think of him for a relative *I* can't
think!—a mullein ?—that we must do without for the
moment ; a monkey flower ?—that we will do without,
altogether ; a lady's slipper ?—say rather a goblin's with
the gout! but, such as the flower-cobbler has made it,
here is one of the kind that people praise, out of the
greenhouse,—and yet a figwort we must have, too ;
which I see on referring to Loudon, may be balm-leaved,
hemp-leaved, tansy-leaved, nettle-leaved, wing-leaved,
heart-leaved, ear-leaved, spear-leaved, or lyre-leaved. I
think I can find a balm-leaved one, though I don't know
what to make of it when I've got it, but it's called a

'Scorodonia' in Sowerby, and something very ugly besides;—I'll put a bit of Teucrium Scorodonia in, to
finish : and now—how will my young Proserpina arrange
her bouquet, and rank the family relations to their con-
tentment ?

5. She has only one kind of flowers in her hand, as
botanical classification stands at present ; and whether
the system be more rational, or in any human sense more
scientific, which puts calceolaria and speedwell together,
—and foxglove and euphrasy ; and runs them on one
side into the mints, and on the other into the nightshades ;—naming them, meanwhile, some from diseases,
some from vermin, some from blockheads, and the rest
anyhow : — or the method I am pleading for, which
teaches us, watchful of their seasonable return and chosen
abiding places, to associate in our memory the flowers
which truly resemble, or fondly companion, or, in time
kept by the signs of Heaven, succeed, each other ; and
to name them in some historical connection with the
loveliest fancies and most helpful faiths of the ancestral
world—Proserpina be judge ; with every maid that sets
flowers on brow or breast—from Thule to Sicily.

6. We will unbind our bouquet, then, and putting all
the rest of its flowers aside, examine the range and nature
of the little blue cluster only.

And first—we have to note of it, that the plan of the
blossom in all the kinds is the same ; an irregular quatrefoil : and irregular quatrefoils are of extreme rarity in

flower form. I don't myself know *one*, except the Veronica. The cruciform vegetables—the heaths, the olives, the lilacs, the little Tormentillas, and the poppies, are all perfectly symmetrical. Two of the petals, indeed, as a rule, are different from the other two, except in the heaths; and thus a distinctly crosslet form obtained, but always an equally balanced one: while in the Veronica, as in the Violet, the blossom always refers itself to a supposed place on the stalk with respect to the ground; and the upper petal is always the largest.

The supposed place is often very suppositious indeed —for clusters of the common veronicas, if luxuriant, throw their blossoms about anywhere. But the idea of an upper and lower petal is always kept in the flower's little mind.

7. In the second place, it is a quite open and flat quatrefoil—so separating itself from the belled quadrature of the heath, and the tubed and primrose-like quadrature of the cruciferæ; and, both as a quatrefoil, and as an open one, it is separated from the foxgloves and snapdragons, which are neither quatrefoils, nor open; but are cinqfoils shut up!

8. In the third place, open and flat though the flower be, it is monopetalous; all the four arms of the cross strictly becoming one in the centre; so that, though the blue foils *look* no less sharply separate than those of a buttercup or a cistus; and are so delicate that one expects them to fall from their stalk if we breathe too

5

near,—do but lay hold of one,—and, at the touch, the
entire blossom is lifted from its stalk, and may be laid,
in perfect shape, on our paper before us, as easily as if
it had been a nicely made-up blue bonnet, lifted off its
stand by the milliner.

I pause here, to consider a little; because I find myself
mixing up two characteristics which have nothing necessary in their relation;—namely, the unity of the blossom,
and its coming easily off the stalk. The separate petals of
the cistus and cherry fall as easily as the foxglove drops
its bells;—on the other hand, there are monopetalous
things that don't drop, but hold on like the convoluta,*
and make the rest of the tree sad for their dying. I do
not see my way to any systematic noting of decadent or
persistent corolla; but, in passing, we may thank the
veronica for never allowing us to see how it fades,† and
being always cheerful and lovely, while it is with us.

9. And for a farther specialty, I think we should take
note of the purity and simplicity of its *floral* blue, not

---

* I find much more difficulty, myself, being old, in using my altered names for species than my young scholars will. In watching
the bells of the purple bindweed fade at evening, let them learn the
fourth verse of the prayer of Hezekiah, as it is in the Vulgate—
" Generatio mea ablata est, et convoluta est a me, sicut tabernaculum
pastoris,"—and they will not forget the name of the fast-fading—ever
renewed—" belle d'un jour."

† " It is Miss Cobbe, I think, who says ' all wild flowers know how
to die gracefully.' "—A.

sprinkling itself with unwholesome sugar like a larkspur, nor varying into coppery or turquoise-like hue as the forget-me-not; but keeping itself as modest as a blue print, pale, in the most frequent kinds; but pure exceedingly; and rejoicing in fellowship with the grey of its native rocks. The palest of all I think it will be well to remember as Veronica Clara, the "Poor Clare" of Veronicas. I find this note on it in my diary,—

'The flower of an exquisite grey-white, like lichen, or shaded hoar-frost, or dead silver; making the long-weathered stones it grew upon perfect with a finished modesty of paleness, as if the flower *could* be blue, and would not, for their sake. Laying its fine small leaves along in embroidery, like Anagallis tenella,—indescribable in the tender feebleness of it—afterwards as it grew, dropping the little blossoms from the base of the spire, before the buds at the top had blown. Gathered, it was happy beside me, with a little water under a stone, and put out one pale blossom after another, day by day.'

10. Lastly, and for a high worthiness, in my estimate, note that it is *wild*, of the wildest, and proud in pure descent of race; submitting itself to no follies of the cur-breeding florist. Its species, though many resembling each other, are severally constant in aspect, and easily recognizable; and I have never seen it provoked to glare into any gigantic impudence at a flower show. Fortunately, perhaps, it is scentless, and so despised.

11. Before I attempt arranging its families, we must

note that while the corolla itself is one of the most constant in form, and so distinct from all other blossoms
that it may be always known at a glance; the leaves and
habit of growth vary so greatly in families of different
climates, and those born for special situations, moist or
dry, and the like, that it is quite impossible to characterize Veronic, or Veronique, vegetation in general terms.
One can say, comfortably, of a strawberry, that it is a
creeper, without expecting at the next moment to see a
steeple of strawberry blossoms rise to contradict us;—we
can venture to say of a foxglove that it grows in a spire,
without any danger of finding, farther on, a carpet of
prostrate and entangling digitalis; and we may pro-
nounce of a buttercup that it grows mostly in meadows,
without fear of finding ourselves, at the edge of the next
thicket, under the shadow of a buttercup-bush growing
into valuable timber. But the Veronica reclines with
the lowly,* upon occasion, and aspires, with the proud;
is here the pleased companion of the ground-ivies, and
there the unrebuked rival of the larkspurs : on the rocks
of Coniston it effaces itself almost into the film of a
lichen ; it pierces the snows of Iceland with the gentian :
and in the Falkland Islands is a white-blossomed evergreen, of which botanists are in dispute whether it be
Veronica or Olive.

---

* See distinction between recumbent and rampant herbs, below,
under ' Veronica Agrestis,' p. 72.

12. Of these many and various forms, I find the manners and customs alike inconstant; and this of especially singular in them—that the Alpine and northern species bloom hardily in contest with the retiring snows, while with us they wait till the spring is past, and offer themselves to us only in consolation for the vanished violet and primrose. As we farther examine the ways of plants, I suppose we shall find some that determine upon a fixed season, and will bloom methodically in June or July, whether in Abyssinia or Greenland; and others, like the violet and crocus, which are flowers of the spring, at whatever time of the favouring or frowning year the spring returns to their country. I suppose also that botanists and gardeners know all these matters thoroughly: but they don't put them into their books, and the clear notions of them only come to me now, as I think and watch.

13. Broadly, however, the families of the Veronica fall into three main divisions,—those which have round leaves lobed at the edge, like ground ivy; those which have small thyme-like leaves; and those which have long leaves like a foxglove's, only smaller—never more than two or two and a half inches long. I therefore take them in these connections, though without any bar between the groups; only separating the Regina from the other thyme-leaved ones, to give her due precedence; and the rest will then arrange themselves into twenty families, easily distinguishable and memorable.

I have chosen for Veronica Regina, the brave Icelandic one, which pierces the snow in first spring, with lovely small shoots of perfectly set leaves, no larger than

FIG. IV.

a grain of wheat; the flowers in a lifted cluster of five or six together, not crowded, yet not loose; large, for veronica—about the size of a silver penny, or say half an inch across—deep blue, with ruby centre.

My woodcut, Fig. 4, is outlined * from the beautiful engraving D. 342,†—there called 'fruticulosa,' from the number of the young shoots.

14. Beneath the Regina, come the twenty easily distinguished families, namely :—

1. Chamædrys. 'Ground-oak.' I cannot tell why so called—its small and rounded leaves having nothing like oak leaves about them, except the serration, which is common to half, at least, of all leaves that grow. But

---

* 'Abstracted' rather, I should have said, and with perfect skill, by Mr. Collingwood (the joint translator of Xenophon's Economics for the 'Bibliotheca Pastorum'). So also the next following cut, Fig. 5.

† Of the references, henceforward necessary to the books I have used as authorities, the reader will please note the following abbreviations :—

C. Curtis's Magazine of Botany.

D. Flora Danica.

F. Figuier.

G. Sibthorpe's Flora Græca.

L. Linnæus. Systema Naturæ.

L. S. Linnæus's Flora Suecica. But till we are quite used to the other letters, I print this reference in words.

L. N. William Curtis's Flora Londinensis. Of the exquisite plates engraved for this book by James Sowerby, note is taken in the close of next chapter.

O. Sowerby's English Wild Flowers; the old edition in thirty-two thin volumes—far the best.

S. Sowerby's English Wild Flowers; the modern edition in ten volumes.

the idea is all over Europe, apparently. Fr. 'petit chêne:' German and English 'Germander,' a merely corrupt form of Chamædrys.

The representative English veronica "Germander Speedwell"—very prettily drawn in S. 986 ; too tall and weed-like in D. 448.

2. Hederifolia. Ivy-leaved: but more properly, cymbalaria-leaved. It is the English field representative, though blue-flowered, of the Byzantine white veronica, V. Cymbalaria, very beautifully drawn in G. 9. Hederifolia well in D. 428.

3. Agrestis. Fr. 'Rustique.' We ought however clearly to understand whether 'agrestis,' used by English botanists, is meant to imply a literally field flower, or only a 'rustic' one, which might as properly grow in a wood. I shall always myself use 'agrestis' in the literal sense, and 'rustica' for 'rustique.' I see no reason, in the present case, for separating the Polite from the Rustic flower: the agrestis, D. 449 and S. 971, seems to me not more meekly recumbent, nor more frankly cultureless, than the so-called Polita, S. 972 : there seems also no French acknowledgment of its politeness, and the Greek family, G. 8, seem the rudest and wildest of all.

Quite a *field* flower it is, I believe, lying always low on the ground ; recumbent, but not creeping. Note this difference: no fastening roots are thrown out by the reposing stems of this Veronica ; a creeping or accurately 'rampant' plant roots itself in advancing. Conf. Nos. 5, 6.

4. Arvensis. We have yet to note a still finer distinction in epithet. 'Agrestis' will properly mean a flower of the open ground—yet not caring whether the piece of earth be cultivated or not, so long as it is under clear sky. But when *agri*-culture has turned the unfruitful acres into 'arva beata,'—if then the plant thrust itself between the furrows of the plough, it is properly called 'Arvensis.'

I don't quite see my way to the same distinction in English,—perhaps I may get into the habit, as time goes on, of calling the Arvenses consistently furrow-flowers, and the Agrestes field-flowers. Furrow-veronica is a tiresomely long name, but must do for the present, as the best interpretation of its Latin character, " vulgatissima in cultis et arvis." D. 515. The blossom itself is exquisitely delicate ; and we may be thankful, both here and in Denmark, for such a lovely 'vulgate.'

5. Montana. D. 1201. The first really creeping plant we have had to notice. It throws out roots from the recumbent stems. Otherwise like agrestis, it has leaves like ground-ivy. Called a wood species in the text of D.

6. Persica. An eastern form, but now perfectly naturalized here—D. 1982; S. 973. The flowers very large, and extremely beautiful, but only one springing from each leaf-axil.

Leaves and stem like Montana ; and also creeping with new roots at intervals.

7. Triphylla, (not triphyll*os*,—see Flora Suecica, 22).
Meaning trifid-leaved ; but the leaf is really divided into
five lobes, not three—see S. 974, and G. 10. The palmate
form of the leaf seems a mere caprice, and indicates no
transitional form in the plant: it may be accepted as only
a momentary compliment of mimicry to the geraniums.
The Siberian variety, 'multifida,' C. 1679, divides itself
almost as the submerged leaves of the water-ranunculus.

The triphylla itself is widely diffused, growing alike
on the sandy fields of Kent, and of Troy. In D. 627 is
given an extremely delicate and minute northern type,
the flowers springing as in Persica, one from each leaf-
axil, and at distant intervals.

8. Officinalis. D. 248, S. 294. Fr. 'Veronique offi-
cinale'; (Germ. Gebrauchlicher Ehrenpreis,) our commonest English and Welsh speedwell; richest in cluster
and frankest in roadside growth, whether on bank or
rock; but assuredly liking *either* a bank *or* a rock, and
the top of a wall better than the shelter of one. Un-
countable ·'myriads,' I am tempted to write, but, cau-
tiously and literally, 'hundreds' of blossoms—if one
could count,—ranging certainly towards the thousand in
some groups, all bright at once, make our Westmoreland
lanes look as if they were decked for weddings, in early
summer. In the Danish Flora it is drawn small and
poor; its southern type being the true one : but it is difficult to explain the difference between the look of a
flower which really *suffers*, as in this instance, by a colder

climate, and becomes mean and weak, as well as dwarfed;
and one which is braced and brightened by the cold,
though diminished, as if under the charge and charm of
an affectionate fairy, and becomes a joyfully patriotic in-
heritor of wilder scenes and skies. Medicinal, to soul
and body alike, this gracious and domestic flower; though
astringent and bitter in the juice. It is the Welsh deeply
honoured ' Fluellen.'—See final note on the myth of
Veronica, see § 18.

9. Thymifolia. Thyme-leaved, G. 6. Of course the
longest possible word—serpyllifolia—is used in S. 978.
It is a high mountain plant, growing on the top of Crete
as the snow retires; and the Veronica minor of Gerarde;
"the roote is small and threddie, taking hold of the *upper surface* of the earth, where it spreadeth." So also
it is drawn as a creeper in F. 492, where the flower appears to be oppressed and concealed by the leafage.

10. Minuta, called 'hirsuta' in S. 985: an ugly characteristic to name the lovely little thing by. The distinct blue lines in the petals might perhaps justify 'picta'
or 'lineata,' rather than an epithet of size; but I suppose
it is Gerarde's Minima, and so leave it, more safely named
as 'minute' than 'least.' For I think the next variety
may dispute the leastness.

11. Verna. D. 252. Mountains, in dry places in
early spring. Upright, and confused in the leafage,
which is sharp-pointed and close set, much hiding the
blossom, but of extreme elegance, fit for a sacred ˴ fore-

ground; as any gentle student will feel, who copies this
outline from the Flora Danica, Fig. 5.

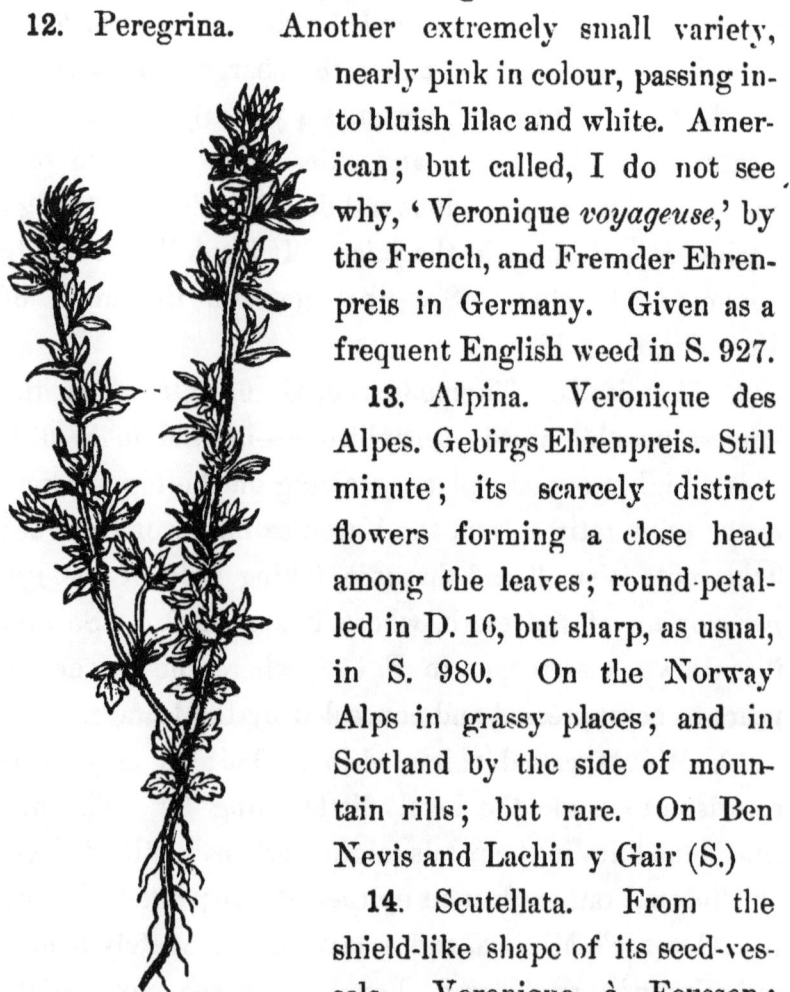

12. Peregrina. Another extremely small variety,
nearly pink in colour, passing in-
to bluish lilac and white. Amer-
ican; but called, I do not see
why, 'Veronique *voyageuse*,' by
the French, and Fremder Ehrenpreis in Germany. Given as a
frequent English weed in S. 927.

13. Alpina. Veronique des
Alpes. Gebirgs Ehrenpreis. Still
minute; its scarcely distinct
flowers forming a close head
among the leaves; round-petal-
led in D. 16, but sharp, as usual,
in S. 980. On the Norway
Alps in grassy places; and in
Scotland by the side of mountain rills; but rare. On Ben
Nevis and Lachin y Gair (S.)

14. Scutellata. From the
shield-like shape of its seed-vessels. Veronique à Ecusson;
Schildfruchtiger Ehrenpreis.

FIG. V.

But the seed-vessels are more heart shape than shield.
Marsh Speedwell. S. 988, D. 209,—in the one pink,
in the other blue; but again in D..1561, pink.

"In flooded meadows, common." (D.) A spoiled
and scattered form; the seeds too conspicuous, but the
flowers very delicate, hence 'Gratiola minima' in Gesner.
The confused ramification of the clusters worth noting,
in relation to the equally straggling fibres of root.

15. Spicata. S. 982 : very prettily done, representing
the inside of the flower as deep blue, the outside pale.
The top of the spire, all calices, the calyx being indeed,
through all the veronicas, an important and persistent
member.

The tendency to arrange itself in spikes is to be noted
as a degradation of the veronic character; connecting it
on one side with the snapdragons, on the other with the
ophryds. In Veronica Ophrydea, (C. 2210,) this resemblance to the contorted tribe is carried so far that "the
corolla of the veronica becomes irregular, the tube gib-
bous, the faux (throat) hairy, and three of the laciniæ
(lobes of petals) variously twisted." The spire of blossom,
violet-coloured, is then close set, and exactly resembles an
ophryd, except in being sharper at the top. The engraved outline of the blossom is good, and very curious.

16. Gentianoides. This is the most directly and curiously imitative among the—shall we call them—'his-
trionic' types of Veronica. It grows exactly like a clustered upright gentian; has the same kind of leaves at its
root, and springs with the same bright vitality among
the retiring snows of the Bithynian Olympus. (G. 5.)
If, however, the Caucasian flower, C. 1002, be the same,

it has lost its perfect grace in luxuriance, growing as large as an asphodel, and with root-leaves half a foot long.

The petals are much veined; and this, of all veronicas, has the lower petal smallest in proportion to the three above,—"triplò aut quadruplò minori." (G.)

17. Stagnarum. Marsh - Veronica. The last four families we have been examining vary from the typical Veronicas not only in their lance-shaped clusters, but in their lengthened, and often every way much enlarged leaves also: and the two which we now will take in association, 17 and 18, carry the change in aspect farthest of any, being both of them true water-plants, with strong stems and thick leaves. The present name of my Veronica Stagnarum is however V. anagallis, a mere insult to the little water primula, which one plant of the Veronica would make fifty of. This is a rank water-weed, having confused bunches of blossom and seed, like unripe currants, dangling from the leaf-axils. So that where the little triphylla, (No. 7, above,) has only one blossom, daintily set, and well seen, this has a litter of twenty-five or thirty on a long stalk, of which only three or four are well out as flowers, and the rest are mere knobs of bud or seed. The stalk is thick (half an inch round at the bottom), the leaves long and misshapen. "Frequens in fossis," D. 203. French, Mouron d'Eau, but I don't know the root or exact meaning of Mouron.

An ugly Australian species, 'labiata,' C. 1660, has leaves two inches long, of the shape of an aloe's, and partly aloeine in texture, "sawed with unequal, fleshy, pointed teeth."

18. Fontium. Brook - Veronica. Brook - *Lime*, the Anglo-Saxon 'lime' from Latin limus, meaning the soft mud of streams. German 'Bach-bunge' (Brook-purse?) ridiculously changed by the botanists into 'Beccabunga,' for a Latin name! Very beautiful in its crowded green leaves as a stream-companion; rich and bright more than watercress. See notice of it at Matlock, in 'Modern Painters,' vol. v.

19. Clara. Veronique des rochers. Saxatilis, I suppose, in Sowerby, but am not sure of having identified that with my own favourite, for which I therefore keep the name 'Clara,' (see above, § 9); and the other rock variety, if indeed another, must be remembered, together with it.

20. Glauca. G. 7. And this, at all events, with the Clara, is to be remembered as closing the series of twenty families, acknowledged by Proserpina. It is a beautiful low-growing ivy-leaved type, with flowers of subdued lilac blue. On Mount Hymettus: no other locality given in the Flora Græca.

15. I am sorry, and shall always be so, when the varieties of any flower which I have to commend to the student's memory, exceed ten or twelve in number; but I am content to gratify his pride with lengthier task, if indeed he will resign himself to the imperative close of the more inclusive catalogue, and be content to know

the twelve, or sixteen, or twenty, acknowledged families, thoroughly; and only in their illustration to think of rarer forms. The object of 'Proserpina' is to make him happily cognizant of the common aspect of Greek and English flowers; under the term 'English,' comprehending the Saxon, Celtic, Norman, and Danish Floras. Of the evergreen shrub alluded to in § 11 above, the Veronica Decussata of the Pacific, which is "a bushy evergreen, with beautifully set cross-leaves, and white blossoms scented like olea fragrans," I should like him only to read with much surprise, and some incredulity, in Pinkerton's or other entertaining travellers' voyages.

16. And of the families given, he is to note for the common simple characteristic, that they are quatrefoils referred to a more or less elevated position on a central stem, and having, in that relation, the lowermost petal diminished, contrary to the almost universal habit of other flowers to develope in such a position the lower petal chiefly, that it may have its full share of light. You will find nothing but blunder and embarrassment result from any endeavour to enter into further particulars, such as "the relation of the dissepiment with respect to the valves of the capsule," etc., etc., since "in the various species of Veronica almost every kind of dehiscence may be observed" (C. under V. perfoliata, 1936, an Australian species). Sibthorpe gives the entire definition of Veronica with only one epithet added to mine, "Corolla quadrifida, *rotata*, laciniâ infimâ angustiore,"

but I do not know what 'rotata' here means, as there is
no appearance of revolved action in the petals, so far as
I can see.

17. Of the mythic or poetic significance of the veronica, there is less to be said than of its natural beauty.
I have not been able to discover with what feeling, or at
what time, its sacred name was originally given ; and the
legend of S. Veronica herself is, in the substance of it,
irrational, and therefore incredible. The meaning of
the term ' rational,' as applied to a legend or miracle, is,
that there has been an intelligible need for the permis-
sion of the miracle at the time when it is recorded ; and
that the nature and manner of the act itself should be
comprehensible in the scope. There was thus quite simple need for Christ to feed the multitudes, and to appear
to S. Paul ; but no need, so far as human intelligence
can reach, for the reflection of His features upon a piece
of linen which could be seen by not one in a million of
the disciples to whom He might more easily, at any time,
manifest Himself personally and perfectly. Nor, I believe, has the story of S. Veronica ever been asserted to
be other than symbolic by the sincere teachers of the
Church ; and, even so far as in that merely explanatory
function, it became the seal of an extreme sorrow, it is
not easy to understand how the pensive fable was associated with a flower so familiar, so bright, and so popu-
larly of good omen, as the Speedwell.

18. Yet, the fact being actually so, and this consecra-

6

tion of the veronica being certainly far more ancient and
earnest than the faintly romantic and extremely absurd
legend of the forget-me-not; the speedwell has assuredly
the higher claim to be given and accepted as a token of
pure and faithful love, and to be trusted as a sweet sign
that the innocence of affection is indeed more frequent,
and the appointed destiny of its faith more fortunate,
than our inattentive hearts have hitherto discerned.

19. And this the more, because the recognized virtues
and uses of the plant are real and manifold; and the
ideas of a peculiar honourableness and worth of life connected with it by the German popular name 'Honour-
prize'; while to the heart of the British race, the same
thought is brought home by Shakespeare's adoption of
the flower's Welsh name, for the faithfullest common
soldier of his ideal king. As a lover's pledge, therefore,
it does not merely mean memory;—for, indeed, why
should love be thought of as such at all, if it need to
promise not to forget?—but the blossom is significant
also of the lover's best virtues, patience in suffering,
purity in thought, gaiety in courage, and serenity in
truth: and therefore I make it, worthily, the clasping
and central flower of the Cytherides.

CHAPTER IV.

GIULIETTA.

1. Supposing that, in early life, one had the power of living to one's fancy,—and why should we not, if the said fancy were restrained by the knowledge of the two great laws concerning our nature, that happiness is increased, not by the enlargement of the possessions, but of the heart; and days lengthened, not by the crowding of emotions, but the economy of them?—if thus taught, we had, I repeat, the ordering of our house and estate in our own hands, I believe no manner of temperance in pleasure would be better rewarded than that of making our gardens gay only with common flowers; and leaving those which needed care for their transplanted life to be found in their native places when we travelled. So long as I had crocus and daisy in the spring, roses in the summer, and hollyhocks and pinks in the autumn, I used to be myself independent of farther horticulture,—and it is only now that I am old, and since pleasant travelling has become impossible to me, that I am thankful to have the white narcissus in my borders, instead of waiting to walk through the fragrance of the meadows of Clarens; and pleased to see the milkwort blue on my scythe-mown

banks, since I cannot gather it any more on the rocks of the Vosges, or in the divine glens of Jura.

2. Among the losses, all the more fatal in being un-felt, brought upon us by the fury and vulgarity of modern life, I count for one of the saddest, the loss of the wish to gather a flower in travelling. The other day, —whether indeed a sign of some dawning of doubt and remorse in the public mind, as to the perfect jubilee of railroad journey, or merely a piece of the common daily flattery on which the power of the British press first de-pends, I cannot judge;—but, for one or other of such motives, I saw lately in some illustrated paper, a pictorial comparison of old-fashioned and modern travel, represent-ing, as the type of things passed away, the outside passen-gers of the mail shrinking into huddled and silent dis-tress from the swirl of a winter snowstorm; and for type of the present Elysian dispensation, the inside of a first-class saloon carriage, with a beautiful young lady in the last pattern of Parisian travelling dress, conversing, Daily news in hand, with a young officer—her fortunate vis-à-vis — on the subject of our military successes in Afghanistan and Zululand.*

3. I will not, in presenting—it must not be called the other side, but the supplementary, and wilfully omitted, facts, of this ideal,—oppose, as I fairly might, the dis-

---

* See letter on the last results of our African campaigns, in the *Morning Post* of April 14th, of this year.

comforts of a modern cheap excursion train, to the
chariot-and-four, with outriders and courier, of ancient
noblesse. I will compare only the actual facts, in the
former and in latter years, of my own journey from Paris
to Geneva. As matters are now arranged, I find myself,
at half past eight in the evening, waiting in a confused
crowd with which I am presently to contend for a seat,
in the dim light and cigar-stench of the great station of
the Lyons line. Making slow way through the hostili-
ties of the platform, in partly real, partly weak politeness, as may be, I find the corner seats of course already
full of prohibitory cloaks and umbrellas; but manage to
get a middle back one; the net overhead is already sur-
charged with a bulging extra portmanteau, so that I
squeeze my desk as well as I can between my legs, and
arrange what wraps I have about my knees and shoulders.
Follow a couple of hours of simple patience, with nothing to entertain one's thoughts but the steady roar of the
line under the wheels, the blinking and dripping of the
oil lantern, and the more or less ungainly wretchedness,
and variously sullen compromises and encroachments of
posture, among the five other passengers preparing themselves for sleep: the last arrangement for the night being to shut up both windows, in order to effect, with our
six breaths, a salutary modification of the night air.

4. The banging and bumping of the carriages over the
turn-tables wakes me up as I am beginning to doze, at
Fontainebleau, and again at Sens; and the trilling and

thrilling of the little telegraph bell establishes itself in
my ears, and stays there, trilling me at last into a shiver-
ing, suspicious sort of sleep, which, with a few vaguely
fretful shrugs and fidgets, carries me as far as Tonnerre,
where the 'quinze minutes d'arret' revolutionize everything; and I get a turn or two on the platform, and
perhaps a glimpse of the stars, with promise of a clear
morning; and so generally keep awake past Mont Bard,
remembering the happy walks one used to have on the
terrace under Buffon's tower, and thence watching, if
perchance, from the mouth of the high tunnel, any film
of moonlight may show the far undulating masses of the
hills of Citeaux. But most likely one knows the place
where the great old view used to be only by the sensible
quickening of the pace as the train turns down the incline, and crashes through the trenched cliffs into the confusion and high clattering vault of the station at Dijon.

5. And as my journey is almost always in the springtime, the twisted spire of the cathedral usually shows itself against the first grey of dawn, as we run out again
southwards: and resolving to watch the sunrise, I fall
more complacently asleep,—and the sun is really up by
the time one has to change carriages, and get morning
coffee at Macon. And from Amberieux, through the
Jura valley, one is more or less feverishly happy and
thankful, not so much for being in sight of Mont Blanc
again, as in having got through the nasty and gloomy
night journey; and then the sight of the Rhone and

the Salève seems only like a dream, presently to end in nothingness; till, covered with dust, and feeling as if one never should be fit for anything any more, one staggers down the hill to the Hotel des Bergues, and sees the dirtied Rhone, with its new iron bridge, and the smoke of a new factory exactly dividing the line of the aiguilles of Chamouni.

6. That is the journey as it is now,—and as, for me, it must be; except on foot, since there is now no other way of making it. But this *was* the way we used to manage it in old days :—

Very early in Continental transits we had found out that the family travelling carriage, taking much time and ingenuity to load, needing at the least three, usually four—horses, and on Alpine passes six, not only jolted and lagged painfully on bad roads, but was liable in every way to more awkward discomfitures than lighter vehicles; getting itself jammed in archways, wrenched with damage out of ruts, and involved in volleys of justifiable reprobation among market stalls. So when we knew better, my father and mother always had their own old-fashioned light two-horse carriage to themselves, and I had one made with any quantity of front and side pockets for books and picked up stones; and hung very low, with a fixed side-step, which I could get off or on with the horses at the trot; and at any rise or fall of the road, relieve them, and get my own walk, without troubling the driver to think of me.

7. Thus, leaving Paris in the bright spring morning, when the Seine glittered gaily at Charenton, and the arbres de Judée were mere pyramids of purple bloom round Villeneuve-St.-Georges, one had an afternoon walk among the rocks of Fontainebleau, and next day we got early into Sens, for new lessons in its cathedral aisles, and the first saunter among the budding vines of the coteaux. I finished my plate of the Tower of Giotto, for the 'Seven Lamps,' in the old inn at Sens, which Dickens has described in his wholly matchless way in the last chapter of 'Mrs. Lirriper's Lodgings'. The next day brought us to the oolite limestones at Mont Bard, and we always spent the Sunday at the Bell in Dijon. Monday, the drive of drives, through the village of Genlis, the fortress of Auxonne, and up the hill to the vine-surrounded town of Dole; whence, behold at last the limitless ranges of Jura, south and north, beyond the woody plain, and above them the 'Derniers Rochers' and the white square-set summit, worshipped ever anew. Then at Poligny, the same afternoon, we gathered the first milkwort for that year; and on Tuesday, at St. Laurent, the wild lily of the valley; and on Wednesday, at Morez, gentians.

And on Thursday, the *eighth or ninth* day from Paris, days all spent patiently and well, one saw from the gained height of Jura, the great Alps unfold themselves in their chains and wreaths of incredible crest and cloud.

8. Unhappily, during all the earliest and usefullest

years of such travelling, I had no thought of ever taking
up botany as a study; feeling well that even geology,
which was antecedent to painting with me, could not be
followed out in connection with art but under strict
limits, and with sore shortcomings. It has only been
the later discovery of the uselessness of old scientific
botany, and the abominableness of new, as an element
of education for youth;—and my certainty that a true
knowledge of their native Flora was meant by Heaven
to be one of the first heart-possessions of every happy
boy and girl in flower-bearing lands, that have compelled
me to gather into system my fading memories, and wandering thoughts.* And of course in the diaries written
at places of which I now want chiefly the details of the
Flora, I find none; and in this instance of the milkwort,
whose name. I was first told by the Chamouni guide,
Joseph Couttet, then walking with me on the unperilous
turf of the first rise of the Vosges, west of Strasburg,
and rebuking me indignantly for my complaint that,
being then thirty-seven years old, and not yet able to
draw the great plain and distant spire, it was of no use
trying in the poor remainder of life to do anything serious,—then, and there, I say, for the first time examining
the strange little flower, and always associating it, since,
with the limestone crags of Alsace and Burgundy, I

---

* I deliberately, not garrulously, allow more autobiography in
'Proserpina' than is becoming, because I know not how far I may be
permitted to carry on that which was begun in 'Fors.'

don't find a single note of its preferences or antipathies
in other districts, and cannot say a word about the soil it
chooses, or the height it ventures, or the familiarities to
which it condescends, on the Alps or Apennines.

9. But one thing I have ascertained of it, lately at
Brantwood, that it is capricious and fastidious beyond
any other little blossom I know of. In laying out the
rock garden, most of the terrace sides were trusted to
remnants of the natural slope, propped by fragments of
stone, among which nearly every other wild flower that
likes sun and air, is glad sometimes to root itself. But
at the top of all, one terrace was brought to mathemati-
cally true level of surface, and slope of side, and turfed
with delicately chosen and adjusted sods, meant to be kept
duly trim by the scythe. And *only* on this terrace does
the Giulietta choose to show herself,—and even there,
not in any consistent places, but gleaming out here in
one year, there in another, like little bits of unexpected
sky through cloud ; and entirely refusing to allow either
bank or terrace to be mown the least trim during *her*
time of disport there. So spared and indulged, there
are no more wayward things in all the woods or wilds;
no more delicate and perfect things to be brought up by
watch through day and night, than her recumbent clusters, trickling, sometimes almost gushing through the
grass, and meeting in tiny pools of flawless blue.

10. I will not attempt at present to arrange the varieties of the Giulietta, for I find that all the larger and

presumably characteristic forms belong to the Cape; and only since Mr. Froude came back from his African explorings have I been able to get any clear idea of the brilliancy and associated infinitude of the Cape flowers. If I could but write down the substance of what he has told me, in the course of a chat or two, which have been among the best privileges of my recent stay in London, (prolonged as it has been by recurrence of illness,) it would be a better summary of what should be generally known in the natural history of southern plants than I could glean from fifty volumes of horticultural botany. In the meantime, everything being again thrown out of gear by the aforesaid illness, I must let this piece of ' Proserpina ' break off, as most of my work does—and as perhaps all of it may soon do—leaving only suggestion for the happier research of the students who trust me thus far.

11. Some essential points respecting the flower I shall note, however, before ending. There is one large and frequent species of it of which the flowers are delicately yellow, touched with tawny red, forming one of the chief elements of wild foreground vegetation in the healthy districts of hard Alpine limestone.* This is, I believe,

---

* In present Botany, Polygala Chamæbuxus; C. 316: or, in English, Much Milk Ground-box. It is not, as matters usually go, a name to be ill thought of, as it really contains three ideas; and the plant does, without doubt, somewhat resemble box, and grows on the ground;—far more fitly called 'ground-box' than the Veronica

the only European type of the large Cape varieties, in all
of which, judging from such plates as have been accessible
to me, the crests or fringes of the lower petal are less
conspicuous than in the smaller species; and the flower
almost takes the aspect of a broom-blossom or pease-
blossom. In the smaller European varieties, the white
fringes of the lower petal are the most important and
characteristic part of the flower, and they are, among
European wild flowers, absolutely without any likeness
of associated structure. The fringes or crests which,
towards the origin of petals, so often give a frosted or
gemmed appearance to the centres of flowers, are here
thrown to the extremity of the petal, and suggest an almost coralline structure of blossom, which in no other
instance whatever has been imitated, still less carried out
into its conceivable varieties of form. How many such
varieties might have been produced if these fringes of the
Giulietta, or those already alluded to of Lucia nivea, had
been repeated and enlarged ; as the type, once adopted
for complex bloom in the thistle-head, is multiplied in
the innumerable gradations of thistle, teasel, hawkweed,
and aster! We might have had flowers edged with lace
finer than was ever woven by mortal fingers, or tasselled

'ground-oak.' I want to find a pretty name for it in connection with
Savoy or Dauphiné, where it indicates, as above stated, the *healthy*
districts of *hard* limestone. I do not remember it as ever occurring
among the dark and moist shales of the inner mountain ranges, which
at once confine and pollute the air.

and braided with fretwork of silver, never tarnished—or
hoarfrost that grew brighter in the sun. But it was not
to be, and after a few hints of what might be done in
this kind, the Fate, or Folly, or, on recent theories, the
extreme fitness—and consequent survival, of the Thistles
and Dandelions, entirely drives the fringed Lucias and
blue-flushing milkworts out of common human neighbourhood, to live recluse lives with the memories of the
abbots of Cluny, and pastors of Piedmont.

12. I have called the Giulietta 'blue-*flushing*' because
it is one of the group of exquisite flowers which at the
time of their own blossoming, breathe their colour into
the surrounding leaves and supporting stem. Very not-
ably the Grape hyacinth and Jura hyacinth, and some
of the Vestals, empurpling all their green leaves even
to the ground: a quite distinct nature in the flower, observe, this possession of a power to kindle the leaf and
stem with its own passion, from that of the heaths, roses,
or lilies, where the determined bracts or calices assert
themselves in opposition to the blossom, as little pine-
leaves, or mosses, or brown-paper packages, and the like.

13. The Giulietta, however, is again entirely separate
from the other leaf-flushing blossoms, in that, after the
two green leaves next the flower have glowed with its
blue, while it lived, they do not fade or waste with it, but
return to their own former green simplicity, and close
over it to protect the seed. I only know this to be the
case with the Giulietta Regina; but suppose it to be

(with variety of course in the colours) a condition in
other species,—though of course nothing is ever said of
it in the botanical accounts of them. I gather, however,
from Curtis's careful drawings that the prevailing colour
of the Cape species is purple, thus justifying still further
my placing them among the Cytherides ; and I am content to take the descriptive epithets at present given
them, for the following five of this southern group, hoping that they may be explained for me afterwards by
helpful friends.

14. Bracteolata, C. 345.

Oppositifolia, C. 492.

Speciosa, C. 1790.

These three all purple, and scarcely distinguishable from
sweet pease-blossom, only smaller.

Stipulacea, C. 1715. Small, and very beautiful, lilac
and purple, with a leaf and mode of growth like rose-
mary. The "Foxtail" milkwort, whose name I don't accept, C. 1006, is intermediate between this and the next
species.

15. Mixta, C. 1714. I don't see what mingling is
meant, except that it is just like Erica tetralix in the
leaf, only, apparently, having little four-petalled pinks
for blossoms. This appearance is thus botanically explained. I do not myself understand the description,
but copy it, thinking it may be of use to somebody.
" The apex of the carina is expanded into a two-lobed
plain petal, the lobes of which are emarginate. This ap-

pendix is of a bright rose colour, and forms the principal
part of the flower." The describer relaxes, or relapses,
into common language so far as to add that ' this appen-
dix ' " dispersed among the green foliage in every part
of the shrub, gives it a pretty lively appearance."

Perhaps this may also be worth extracting.

" Carina, deeply channeled, *of a saturated purple* within, sides folded together, so as to include and firmly
embrace the style and stamens, which, when arrived at
maturity, upon being moved, escape elastically from their
confinement, and strike against the two erect petals or
alæ—by which the pollen is dispersed.

"Stem shrubby, with long flexile branches." (Length
or height not told. I imagine like an ordinary heath's.)

The term ' carina,' occurring twice in the above description, is peculiar to the structure of the pease and milkworts; we will examine it afterwards. The European
varieties of the milkwort, except the chamæbuxus, are
all minute,—and, their ordinary epithets being at least
inoffensive, I give them for reference till we find prettier
ones; altering only the Calcarea, because we could not
have a ' Chalk Juliet,' and two varieties of the Regina,
changed for reason good—her name, according to the
last modern refinements of grace and ease in pronuncia-
tion, being Eu-vularis, var. genuina! My readers may
more happily remember her and her sister as follows :—

16. (I.) Giulietta Regina. Pure blue. The same in colour, form, and size, throughout Europe.

(II.) Giulietta Soror-Reginæ. Pale, reddish-blue or white in the flower, and smaller in the leaf, otherwise like the Regina.

(III.) Giulietta Depressa. The smallest of those I can find drawings of. Flowers, blue; lilac in the fringe, and no bigger than pins' heads; the leaves quite gem-like in minuteness and order.

(IV.) Giulietta Cisterciana. Its present name, 'Calcarea,' is meant, in botanic Latin, to express its growth on limestone or chalk mountains. But we might as well call the South Down sheep, Calcareous mutton. My epithet will rightly associate it with the Burgundian hills round Cluny and Citeaux. Its ground leaves are much larger than those of the Depressa; the flower a little larger, but very pale.

(V.) Giulietta Austriaca. Pink, and very lovely, with bold cluster of ground leaves, but itself minute—almost dwarf. Called 'small bitter milkwort' by S. How far distinct from the next following one, Norwegian, is not told.

The above five kinds are given by Sowerby as British, but I have never found the
Austriaca myself.

(VI.) Giulietta Amara. Norwegian. Very quaint
in blossom outline, like a little blue rabbit
with long ears. D. 1169.

17. Nobody tells me why either this last or No. 5 have
been called bitter; and Gerarde's five kinds are distinguished only by colour—blue, red, white, purple, and
"the dark, of an overworn ill-favoured colour, which
maketh it to differ from all others of his kind." I find
no account of this ill-favoured one elsewhere. The white
is my Soror Reginæ; the red must be the Austriaca; but
the purple and overworn ones are perhaps now overworn
indeed. All of them must have been more common in
Gerarde's time than now, for he goes on to say "Milk-
woort is called *Ambarualis flos,* so called because it doth
specially flourish in the Crosse or Gang-weeke, or Roga-
tion-weeke, of which flowers, the maidens which use in
the countries to walk the procession do make themselves •
garlands and nosegaies, in English we may call it Crosse
flower, Gang flower, Rogation flower, and Milk-woort."

18. Above, at page 197, vol. i., in first arranging the
Cytherides, I too hastily concluded that the ascription to
this plant of helpfulness to nursing mothers was 'more
than ordinarily false'; thinking that its rarity could
never have allowed it to be fairly tried. If indeed true,
or in any degree true, the flower has the best right of all

7

to be classed with the Cytherides, and we might have as much of it for beauty and for service as we choose, if we only took half the pains to garnish our summer gardens with living and life-giving blossom, that we do to garnish our winter gluttonies with dying and useless ones.

19. I have said nothing of root, or fruit, or seed, having never had the hardness of heart to pull up a milkwort cluster—nor the chance of watching one in seed :—The pretty thing vanishes as it comes, like the blue sky of April, and leaves no sign of itself—that *I* ever found. The botanists tell me that its fruit " dehisces loculicidally," which I suppose is botanic for " splits like boxes," (but boxes shouldn't split, and didn't, as we used to make and handle them before railways). Out of the split boxes fall seeds—too few ; and, as aforesaid, the plant never seems to grow again in the same spot. I should thankfully receive any notes from friends happy enough to live near milkwort banks, on the manner of its nativity.

20. Meanwhile, the Thistle, and the Nettle, and the Dock, and the Dandelion are cared for in their generations by the finest arts of—Providence, shall we say ? or of the spirits appointed to punish our own want of Providence ? May I ask the reader to look back to the seventh chapter of the first volume, for it contains suggestions of thoughts which came to me at a time of very earnest and faithful inquiry, set down, I now see too shortly, under the press of reading they involved, but intelligible enough if they are read as slowly as they were

XI.

States of Adversity.

written, and especially note the paragraph of summary of p. 121 on the power of the Earth Mother, as Mother, and as *judge ;* watching and rewarding the conditions which induce adversity and prosperity in the kingdoms of men : comparing with it carefully the close of the fourth chapter, p. 85,* which contains, for the now recklessly multiplying classes of artists and colonists, truths essential to their skill, and inexorable upon their labour.

21. The pen-drawing facsimiled by Mr. Allen with more than his usual care in the frontispiece to this number of 'Proserpina,' was one of many executed during the investigation of the schools of Gothic (German, and later French), which founded their minor ornamentation on the serration of the thistle leaf, as the Greeks on that of the Acanthus, but with a consequent, and often morbid, love of thorny points, and insistance upon jagged or knotted intricacies of stubborn vegetation, which is connected in a deeply mysterious way with the gloomier forms of Catholic asceticism.†

---

* Which, with the following page, is the summary of many chapters of 'Modern Painters :' and of the aims kept in view throughout ' Munera Pulveris.' The three kinds of Desert specified—of Reed, Sand, and Rock—should be kept in mind as exhaustively including the states of the earth neglected by man. For instance of a Reed desert, produced *merely* by his neglect, see Sir Samuel Baker's account of the choking up of the bed of the White Nile. Of the sand desert, Sir F. Palgrave's journey from the Djowf to Hâyel, vol. i., p. 92.

† This subject is first entered on in the ' Seven Lamps,' and carried forward in the final chapters of ' Modern Painters,' to the point where

22. But also, in beginning 'Proserpina,' I intended to give many illustrations of the light and shade of foreground leaves belonging to the nobler groups of thistles, because I thought they had been neglected by ordinary botanical draughtsmen; not knowing at that time either the original drawings at Oxford for the 'Flora Græca,' or the nobly engraved plates executed in the close of the last century for the 'Flora Danica' and 'Flora Londinensis.' The latter is in the most difficult portraiture of the larger plants, even the more wonderful of the two; and had I seen the miracles of skill, patience, and faithful study which are collected in the first and second volumes, published in 1777 and 1798, I believe my own work would never have been undertaken.* Such as it is, however, I may still, health being granted me, persevere in it; for my own leaf and branch studies express conditions of shade which even these most exquisite botanical plates ignore; and exemplify uses of the pen and pencil which cannot be learned from the inimitable fineness of line engraving. The frontispiece to this number, for instance, (a seeding head of the commonest field-thistle of our London suburbs,) copied with a steel pen on smooth grey paper, and the drawing softly touched with

---

I hope to take it up for conclusion, in the sections of 'Our Fathers have told us' devoted to the history of the fourteenth century.

* See in the first volume, the plates of Sonchus Arvensis and Tussilago Petasites; in the second, Carduus tomentosus and Picris Echioides.

white on the nearer thorns, may well surpass the effect of
the plate.

23. In the following number of 'Proserpina' I have
been tempted to follow, with more minute notice than
usual, the 'conditions of adversity' which, as they fret the
thistle tribe into jagged malice, have humbled the beauty
of the great domestic group of the Vestals into confused
likenesses of the Dragonweed and Nettle: but I feel
every hour more and more the necessity of separating
the treatment of subjects in 'Proserpina' from the microscopic curiosities of recent botanic illustration, nor
shall this work close, if my strength hold, without fulfil-
ling in some sort, the effort begun long ago in 'Modern
Painters,' to interpret the grace of the larger blossoming
trees, and the mysteries of leafy form which clothe the
Swiss precipice with gentleness, and colour with softest
azure the rich horizons of England and Italy.

CHAPTER V.

1. IT ought to have been added to the statements of general law in irregular flowers, in Chapter I. of this volume, § 6, that if the petals, while brought into relations of inequality, still retain their perfect petal form,—and whether broad or narrow, extended or reduced, remain clearly *leaves*, as in the pansy, pea, or azalea, and assume no grotesque or obscure outline,—the flower, though injured, is not to be thought of as corrupted or misled. But if any of the petals lose their definite character as such, and become swollen, solidified, stiffened, or strained into any other form or function than that of petals, the flower is to be looked upon as affected by some kind of constant evil influence; and, so far as we conceive of any spiritual power being concerned in the protection or affliction of the inferior orders of creatures, it will be felt to bear the aspect of possession by, or pollution by, a more or less degraded Spirit.*

2. I have already enough spoken of the special mani-

---

* For the sense in which this word is used throughout my writings, see the definition of it in the 52nd paragraph of the ' Queen of the Air,' comparing with respect to its office in plants, §§ 59–60.

festation of this character in the orders Contorta and
Satyrium, vol. i., p. 91, and the reader will find the
parallel aspects of the Draconidæ dwelt upon at length
in the 86th and 87th paragraphs of the 'Queen of the
Air,' where also their relation to the labiate group is
touched upon. But I am far more embarrassed by the
symbolism of that group which I called 'Vestales,' from
their especially domestic character and their serviceable
purity; but which may be, with more convenience perhaps, simply recognizable as ' Menthæ.'

3. These are, to our northern countries, what the spice-
bearing trees are in the tropics;—our thyme, lavender,
mint, marjoram, and their like, separating themselves
not less in the health giving or strengthening character
of their scent from the flowers more or less enervating in
perfume, as the rose, orange, and violet,—than in their
humble colours and forms from the grace and splendour
of those higher tribes; thus allowing themselves to be
summed under the general word ' balm' more truly than
the balsams from which the word is derived. Giving
the most pure and healing powers to the air around
them; with a comfort of warmth also, being mostly in
dry places, and forming sweet carpets and close turf;
but only to be rightly enjoyed in the open air, or indoors
when dried; not tempting any one to luxury, nor ex-
pressive of any kind of exultation. Brides do not deck
themselves with thyme, nor do we wreathe triumphal
arches with mint.

4. It is most notable, also, farther, that none of these flowers have any extreme beauty in colour. The blue sage is the only one of vivid hue at all; and we never think of it as for a moment comparable to the violet or bluebell: thyme is unnoticed beside heath, and many of the other purple varieties of the group are almost dark and sad coloured among the flowers of summer; while, so far from gaining beauty on closer looking, there is scarcely a blossom of them which is not more or less grotesque, even to ugliness, in outline; and so hooded or lappeted as to look at first like some imperfect form of snapdragon: for the most part spotted also, wrinkled as if by old age or decay, cleft or torn, as if by violence, and springing out of calices which, in their clustering spines, embody the general roughness of the plant.

5. I take at once for example, lest the reader should think me unkind or intemperate in my description, a flower very dear and precious to me; and at this time my chief comfort in field walks. For, now, the reign of all the sweet reginas of the spring is over—the reign of the silvia and anemone, of viola and veronica; and at last, and this year abdicated under tyrannous storm,* the reign of the rose. And the last foxglove-bells are nearly fallen; and over all my fields and by the brooksides are coming up the burdock, and the coarse and vainly white aster, and the black knapweeds; and there is only one

---

* Written in 1880.

flower left to love among the grass,—the soft, warm-scented Brunelle.

6. *P*runell, *or* Brunell—Gerarde calls it, and Brunella, rightly and authoritatively, Tournefort; Prunella, carelessly, Linnæus, and idly following him, the moderns, casting out all the meaning and help of its name—of which presently. Selfe-heale, Gerarde and Gray call it, in English—meaning that who has this plant needs no physician.

7. As I look at it, close beside me, it seems as if it would reprove me for what I have just said of the poverty of colour in its tribe; for the most glowing of violets could not be lovelier than each fine purple gleam of its hooded blossoms. But their flush is broken and oppressed by the dark calices out of which they spring, and their utmost power in the field is only of a saddened amethystine lustre, subdued with furry brown. And what is worst in the victory of the darker colour is the disorder of the scattered blossoms ;—of all flowers I know, this is the strangest, in the way that here and there, only in their cluster, its bells rise or remain, and it always looks as if half of them had been shaken off, and the top of the cluster broken short away altogether.

8. We must never lose hold of the principle that every flower is meant to be seen by human creatures with human eyes, as by spiders with spider eyes. But as the painter may sometimes play the spider, and weave a mesh to entrap the heart, so the beholder may play the

spider, when there are meshes to be disentangled that have entrapped his mind. I take my lens, therefore—to the little wonder of a brown wasps' nest with blue-winged wasps in it,—and perceive therewith the following particulars.

9. First, that the blue of the petals is indeed pure and lovely, and a little crystalline in texture; but that the form and setting of them is grotesque beyond all wonder; the two uppermost joined being like an old-fashioned and enormous hood or bonnet, and the lower one projecting far out in the shape of a cup or cauldron, torn deep at the edges into a kind of fringe.

Looking more closely still, I perceive there is a cluster of stiff white hairs, almost bristles, on the top of the hood; for no imaginable purpose of use or decoration—any more than a hearth-brush put for a helmet-crest,—and that, as we put the flower full in front, the lower petal begins to look like some threatening viperine or shark-like jaw, edged with ghastly teeth,—and yet more, that the hollow within begins to suggest a resemblance to an open throat in which there are two projections where the lower petal joins the lateral ones, almost exactly like swollen glands.

I believe it was this resemblance, inevitable to any careful and close observer, which first suggested the use of the plant in throat diseases to physicians; guided, as in those first days of pharmacy, chiefly by imagination. Then the German name for one of the most fatal of

throat affections, Braune, extended itself into the first
name of the plant, Brunelle.

10. The truth of all popular traditions as to the healing power of herbs will be tried impartially as soon as
men again desire to lead healthy lives; but I shall not in
'Proserpina' retain any of the names of their gathered
and dead or distilled substance, but name them always
from the characters of their life. I retain, however, for
this plant its name Brunella, Fr. Brunelle, because we
may ourselves understand it as a derivation from Brune ;
and I bring it here before the reader's attention as giving
him a perfectly instructive general type of the kind of
degradation which takes place in the forms of flowers
under more or less malefic influence, causing distortion
and disguise of their floral structure. Thus it is not the
normal character of a flower petal to have a cluster of
bristles growing out of the middle of it, nor to be jagged at
the edge into the likeness of a fanged fish's jaw, nor to be
swollen or pouted into the likeness of a diseased gland in
an animal's throat. A really uncorrupted flower suggests
none but delightful images, and is like nothing but itself.

11. I find that in the year 1719, Tournefort defined,
with exactitude which has rendered the definition authoritative for all time, the tribe to which this Brownie
flower belongs, constituting them his fourth class, and
describing them in terms even more depreciatingly im-
aginative than any I have ventured to use myself. I
translate the passage (vol. i., p. 177) :—

12. " The name of Labiate flower is given to a single-petaled flower which, beneath, is attenuated into a tube, and above is expanded into a lip, which is either single or double. It is proper to a labiate flower,—first, that it has a one-leaved calyx (ut calycem habeat *unifolium*), for the most part tubulated, or reminding one of a paper hood (cucullum papyraceum) ; and, secondly, that its pistil ripens into a fruit consisting of four seeds, which ripen in the calyx itself, as if in their own seed-vessel, by which a labiate flower is distinguished from a personate one, whose pistil becomes a capsule far divided from the calyx (à calyce longè divisam). And a labiate flower differs from rotate, or bell-shaped flowers, which have four seeds, in that the lips of a labiate flower have a gape like the face of a goblin, or ludicrous mask, emulous of animal form."

13. This class is then divided into four sections.

In the first, the upper lip is helmeted, or hooked— " galeatum est, vel falcatum."

In the second, the upper lip is excavated like a spoon —" cochlearis instar est excavatum."

In the third the upper lip is erect.

And in the fourth there is no upper lip at all.

The reader will, I hope, forgive me for at once rejecting a classification of lipped plants into three classes that have lips, and one that has none, and in which the lips of those that have got any, are like helmets and spoons.

Linnæus, in 1758, grouped the family into two divisions, by the form of the calyx, (five-fold or two-fold), and then went into the wildest confusion in distinction of species,—sometimes by the form of corolla, sometimes by that of calyx, sometimes by that of the filaments, sometimes by that of the stigma, and sometimes by that of the seed. As, for instance, thyme is to be identified by the calyx having hairs in its throat, dead nettle by having bristles in its mouth, lion's tail by having bones in its anthers (antheræ punctis osseis adspersæ), and teucrium by having its upper lip cut in two!

14. St. Hilaire, in 1805, divides again into four sections, but as three of these depend on form of corolla, and the fourth on abortion of stamens, the reader may conclude practically, that logical division of the family is impossible, and that all he can do, or that there is the smallest occasion for his doing, is first to understand the typical structure thoroughly, and then to know a certain number of forms accurately, grouping the others round them at convenient distances; and, finally, to attach to their known forms such simple names as may be utterable by children, and memorable by old people, with more ease and benefit than. the 'Galeopsis Eu-te-trahit,' 'Lamium Galeobdalon,' or 'Scutellaria Galericulata,' and the like, of modern botany. But to do this rightly, I must review and amplify some of my former classification, which it will be advisable to do in a separate chapter.

CHAPTER VI.

1. IT is not a little vexing to me, in looking over the very little I have got done of my planned Systema Proserpinæ, to discover a grave mistake in the specifications of Veronica. It is Veronica chamædrys, not officinalis, which is our proper English Speedwell, and Welsh Fluellen; and all the eighth paragraph, p. 74, properly applies to that. Veronica officinalis is an extremely small flower rising on vertical stems out of recumbent leaves; and the drawing of it in the Flora Danica, which I mistook for a stunted northern state, is quite true of the English species,* except that it does not express the recumbent action of the leaves. The proper representation of ground-leafage has never yet been attempted in any botanical work whatever, and as, in recumbent plants, their grouping and action can only be seen from above, the plates of them should always have a dark and rugged background, not only to indicate the position of the eye, but to relieve the forms of the

---

* The plate of Chamædrys, D. 448, is also quite right, and not 'too tall and weedlike,' as I have called it at p. 72.

leaves as they were intended to be shown. I will try to give some examples in the course of this year.

2. I find also, sorrowfully, that the references are wrong in three, if not more, places in that chapter. S. 971 and 972 should be transposed in p. 72. S. 294 in p. 74 should be 984. D. 407 should be inserted after Peregrina, in p. 76; and 203, in fourth line from bottom of p. 78, should be 903. I wish it were likely that these errors had been corrected by my readers,—the rarity of the Flora Danica making at present my references virtually useless: but I hope in time that our public institutes will possess themselves of copies: still more do I hope that some book of the kind will be undertaken by English artists and engravers, which shall be worthy of our own country.

3. Farther, I get into confusion by not always remembering my own nomenclature, and have allowed 'Gentianoides' to remain, for No. 16, though I banish Gentian. It will be far better to call this eastern mountain species 'Olympica': according to Sibthorpe's localization, "in summâ parte, nive solutâ, montis Olympi Bithyni," and the rather that Curtis's plate above referred to shows it in luxuriance to be liker an asphodel than a gentian.

4. I have also perhaps done wrong in considering Veronica polita and agrestis as only varieties, in No. 3. No author tells me why the first is called polite, but its blue seems more intense than that of **agrestis**; and as it

is above described with attention, vol. i., p. 75, as an example of precision in flower-form, we may as well retain it in our list here.   It will be therefore our twenty-first variety,—it is London's fifty-ninth and last.   He translates ' polita ' simply 'polished,' which is nonsense. I can think of nothing to call it but 'dainty,' and will leave it at present unchristened.

5. Lastly.   I can't think why I omitted V. Humifusa, S. 979, which seems to be quite one of the most beautiful of the family—a mountain flower also, and one which I ought to find here; but hitherto I know only among the mantlings of the ground, V. thymifolia and officinalis.   All these, however, agree in the extreme prettiness and grace of their crowded leafage,—the officinalis, of which the leaves are shown much too coarsely serrated in S. 984, forming carpets of finished embroidery which I have never yet rightly examined, because I mistook them for St. John's wort.   They are of a beautiful pointed oval form, serrated so finely that they seem smooth in distant effect, and covered with equally invisible hairs, which seem to collect towards the edge in the variety Hirsuta, S. 985.

For the present, I should like the reader to group the three flowers, S. 979, 984, 985, under the general name of Humifusa, and to distinguish them by a third epithet, which I allow myself when in difficulties, thus:

V. Humifusa, cærulea, the beautiful blue one, which resembles Spicata.

V. Humifusa, officinalis, and,

V. Humifusa, hirsuta: the last seems to me extremely interesting, and I hope to find it and study it carefully.

By this arrangement we shall have only twenty-one species to remember: the one which chiefly decorates the ground again dividing into the above three.

6. These matters being set right, I pass to the business in hand, which is to define as far as possible the subtle relations between the Veronicas and Draconidæ, and again between these and the tribe at present called labiate. In my classification above, vol. i., p. 200, the Draconidæ include the Nightshades; but this was an oversight. Atropa belongs properly to the following class, Moiridæ; and my Draconids are intended to include only the two great families of Personate and Ringent flowers, which in some degree resemble the head of an animal: the representative one being what we call 'snapdragon,' but the French, careless of its snapping power, calf's muzzle—"Muflier, muflande, or muffle de Veau."—Rousseau, 'Lettres,' p. 19.

7. As I examine his careful and sensible plates of it, I chance also on a bit of his text, which, extremely wise and generally useful, I translate forthwith:—

"I understand, my dear, that one is vexed to take so much trouble without learning the names of the plants one examines; but I confess to you in good faith that it never entered into my plan to spare you this little

8

chagrin.   One pretends that Botany is nothing but a
science of words, which only exercises the memory, and
only teaches how to give plants names.   For me, I know
no rational study which is only a science of words : and
to which of the two, I pray you, shall I grant the name
of botanist,—to him who knows how to spit out a name
or a phrase at the sight of a plant, without knowing anything of its structure, or to him who, knowing that structure very well, is ignorant nevertheless of the very arbi-
trary name that one gives to the plant in such and such a
country?   If we only gave to your children an amusing
occupation, we should miss the best half of our purpose,
which is, in amusing them, to exercise their intelligence
and accustom them to attention.   Before teaching them
to name what they see, let us begin by teaching them to
see it.   *That* science, forgotten in all educations, ought
to form the most important part of theirs.   I can never
repeat it often enough—teach them never to be satisfied
with words, ('se payer de mots') and to hold themselves
as knowing nothing of what has reached no farther than
their memories."

8. Rousseau chooses, to represent his 'Personees,' La
Mufflaude, la Linaire, l'Euphraise, la Pediculaire, la
Crête-de-coq, l'Orobanche, la Cimbalaire, la Velvote, la
Digitale, giving plates of snapdragon, foxglove, and
Madonna-herb, (the Cimbalaire), and therefore including
my entire class of Draconidæ, whether open or close
throated.   But I propose myself to separate from them

the flower which, for the present, I have called Monacha,
but may perhaps find hereafter a better name; this one,
which is the best Latin I can find for a nun of the desert, being given to it because all the resemblance either
to calf or dragon has ceased in its rosy petals, and they
resemble—the lower ones those of the mountain thyme,
and the upper one a softly crimson cowl or hood.

9. This beautiful mountain flower, at present, by the
good grace of botanists, known as Pedicularis, from a
disease which it is supposed to give to sheep, is distinguished from all other Draconidæ by its beautifully
divided leaves : while the flower itself, like, as aforesaid, thyme in the three lower petals, rises in the upper
one quite upright, and terminates in the narrow and
peculiar hood from which I have named it ' Monacha.'

10. Two deeper crimson spots with white centres ani-
mate the colour of the lower petals in our mountain kind
—mountain or morass ;—it is vilely drawn in S. 997
under the name of Sylvatica, translated ' Procumbent ' !
As it is neither a wood flower nor a procumbent one,*
and as its rosy colour is rare among morass flowers, I
shall call it simply Monacha Rosea.

I have not the smallest notion of the meaning of the

---

* " Stems numerous from the crown of the root-stock, de-cum-
bent."—S. The effect of the flower upon the ground is always of
an extremely upright and separate plant, never appearing in clusters,
or in any relation to a central root. My epithet ' rosea ' does not
deny its botanical de- or pro-cumbency.

following sentence in S. :—" Upper lip of corolla not
rostrate, with the margin on each side furnished with a
triangular tooth immediately below the apex, but without any tooth below the middle." Why, or when, a lip
is rostrate, or has any 'tooth below the middle,' I do not
know; but the upper *petal* of the corolla is here a very
close gathered hood, with the style emergent downwards,
and the stamens all hidden and close set within.

In this action of the upper petal, and curve of the
style, the flower resembles the Labiates,* and is the
proper link between them and the Draconidæ. The
capsule is said by S. to be oval-ovoid. As eggs always
are oval, I don't feel farther informed by the epithet.
The capsule and seed both are of entirely indescribable
shapes, with any number of sides—very foxglove-like,
and inordinately large. The seeds of the entire family
are ' ovoid-subtrigonous.'—S.

11. I find only two species given as British by S.,
namely, Sylvatica and Palustris ; but I take first for the
Regina, the beautiful Arctic species D. 1105, Flora
Suecica, 555. Rose-coloured in the stem, pale pink in
the flowers (corollæ pallide incarnatæ), the calices furry
against the cold, whence the present ugly name, Hirsuta.
Only on the highest crests of the Lapland Alps.

(2) Rosea, D. 225, there called Sylvatica, as by S.,
presumably because " in pascuis subhumidis non raræ."

---

* Compare especially Galeopsis Angustifolia, D. 3031.

Beautifully drawn, but, as I have described it, vigorously erect, and with no decumbency whatever in any part of it. Root branched, and enormous in proportion to plant, and I fancy therefore must be good for something if one knew it. But Gerarde, who calls the plant Red Rattle, (it having indeed much in common with the Yellow Rattle), says, "It groweth in moist and moorish meadows; the herbe is not only unprofitable, but likewise hurtful, and an infirmity of the meadows."

(3) Palustris, D. 2055, S. 996—scarcely any likeness between the plates. "Everywhere in the meadows," according to D. I leave the English name, Marsh Monacha, much doubting its being more marshy than others.

12. I take next (4 and 5) two northern species, Lapponica, D. 2, and Grönlandica, D. 1166; the first yellow, the second red, both beautiful. The Lap one has its divided leaves almost united into one lovely spear-shaped single leaf. The Greenland one has its red hood much prolonged in front.

(6) Ramosa, also a Greenland species; yellow, very delicate and beautiful. Three stems from one root, but may be more or fewer, I suppose.

13. (7) Norvegica, a beautifully clustered golden flower, with thick stem, D. 30, the only locality given being the Dovrefeldt. "Alpina" and "Flammea" are the synonyms, but I do not know it on the Alps, and it is no more flame-coloured than a cowslip.

Both the Lapland and Norwegian flowers are drawn with their stems wavy, though upright—a rare and pretty habit of growth.

14. (8) Suecica, D. 26, named awkwardly Sceptrum Carolinum, in honour of Charles XII. It is the largest of all the species drawn in D., and contrasts strikingly with (4) and (5) in the strict uprightness of its stem. The corolla is closed at the extremity, which is red; the body of the flower pale yellow. Grows in marshy and shady woods, near Upsal. Linn., Flora Suecica, 553.

The many-lobed but united leaves, at the root five or six inches long, are irregularly beautiful.

15. These eight species are all I can specify, having no pictures of the others named by London,—eleven, making nineteen altogether, and I wish I could find a twentieth and draw them all, but the reader may be well satisfied if he clearly know these eight. The group they form is an entirely distinct one, exactly intermediate between the Vestals and Draconids, and cannot be rightly attached to either; for it is Draconid in structure and affinity—Vestal in form—and I don't see how to get the connection of the three families rightly expressed without taking the Draconidæ out of the groups belonging to the dark Kora, and placing them next the Vestals, with the Monachæ between; for indeed Linaria and several other Draconid forms are entirely innocent and beautiful, and even the Foxglove never does any real

mischief like hemlock, while decoratively it is one of the most precious of mountain flowers. I find myself also embarrassed by my name of Vestals, because of the masculine groups of Basil and Thymus, and I think it will be better to call them simply Menthæ, and to place them with the other cottage-garden plants not yet classed, taking the easily remembered names Mentha, Monacha, Draconida. This will leave me a blank seventh place among my twelve orders at p. 194, vol. i., which I think I shall fill by taking cyclamen and anagillis out of the Primulaceæ, and making a separate group of them. These retouchings and changes are inevitable in a work confessedly tentative and suggestive only ; but in whatever state of imperfection I may be forced to leave 'Proserpina,' it will assuredly be found, up to the point reached, a better foundation for the knowledge of flowers in the minds of young people than any hitherto adopted system of nomenclature.

16. Taking then this re-arranged group, Mentha, Monacha, and Draconida, as a sufficiently natural and convenient one, I will briefly give the essentially botanical relations of the three families.

Mentha and Monacha agree in being essentially hooded flowers, the upper petal more or less taking the form of a cup, helmet or hood, which conceals the tops of the stamens. Of the three lower petals, the lowest is almost invariably the longest ; it sometimes is itself divided again into two, but may be best thought of as single, and with the two lateral ones, distinguished in the Menthæ as the apron and the side pockets.

Plate XII. represents the most characteristic types of the blossoms of Menthæ, in the profile and front views, all a little magnified. The upper two are white basil, purple spotted—growing here at Brantwood always with two terminal flowers. The two middle figures are the purple-spotted dead nettle, Lamium maculatum ; and the two lower, thyme : but I have not been able to draw these as I wanted, the perspectives of the petals being too difficult, and inexplicable to the eye even in the flowers themselves without continually putting them in changed positions.

17. The Menthæ are in their structure essentially quadrate plants; their stems are square, their leaves opposite, their stamens either four or two, their seeds two-carpeled. But their calices are five-sepaled, falling into divisions of two and three ; and the flowers, though essentially four-petaled, may divide either the upper or lower petal, or both, into two lobes, and so present a six-lobed outline. The entire plants, but chiefly the leaves, are nearly always fragrant, and always innocent. None of them sting, none prick, and none poison.

18. The Draconids, easily recognizable by their aspect, are botanically indefinable with any clearness or simplicity. The calyx may be five- or four-sepaled ; the corolla, five- or four-lobed ; the stamens may be two, four, four with a rudimentary fifth, or five with the two

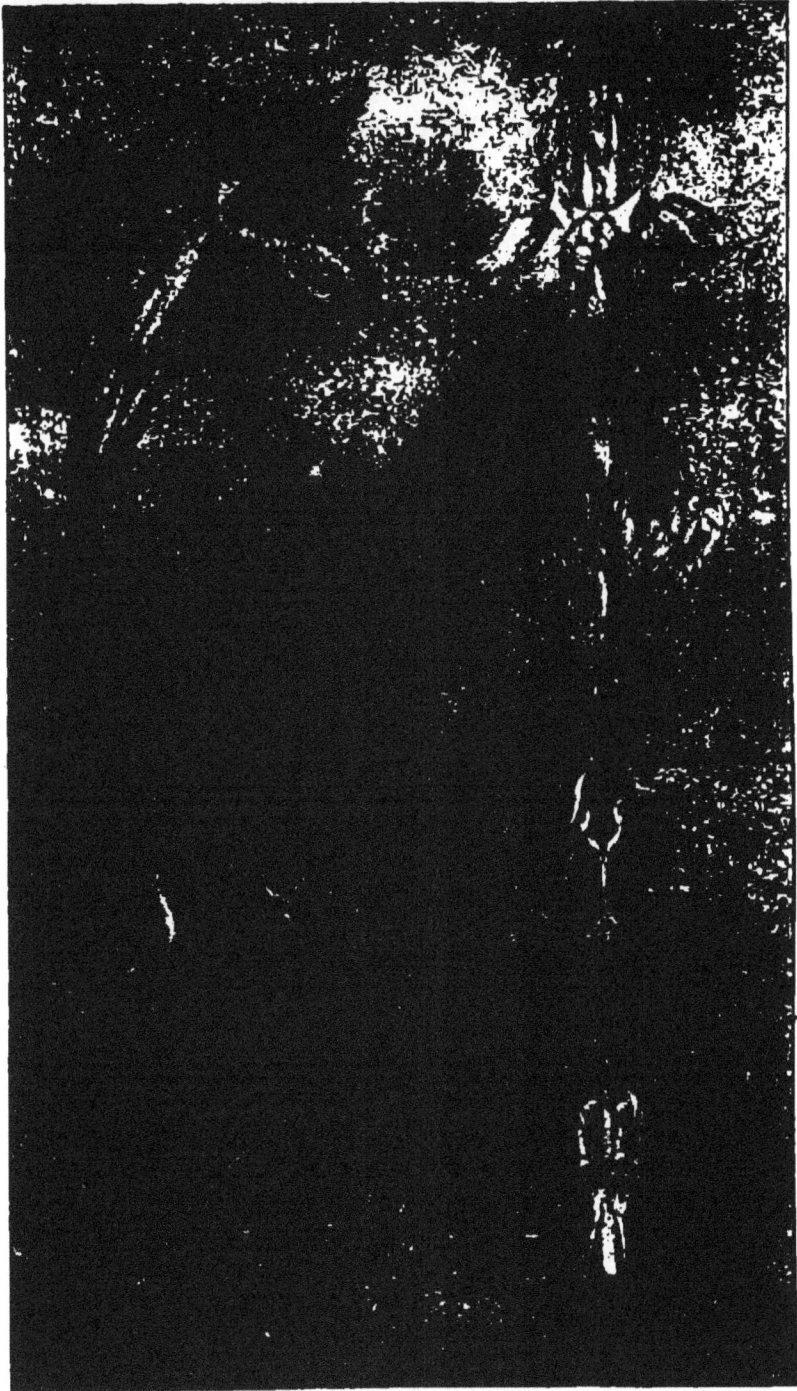

XII.

MENTHÆ.

Profile and Front View of Blossoms (enlarged).

anterior ones longer than the other three! The capsule
may open by two, three, or four valves,—or by pores;
the seeds, generally numerous, are sometimes solitary,
and the leaves may be alternate, opposite, or verticillate.

19. Thus licentious in structure, they are also doubtful in disposition. None that I know of are fragrant,
few useful, many more or less malignant, and some parasitic. The following piece of a friend's letter almost
makes me regret my rescue of them from the dark kingdom of Kora :—

". . . And I find that the Monacha Rosea (Red Rattle is its name,
besides the ugly one) is a perennial, and several of the other draconidæ, foxglove, etc., are biennials, born this year, flowering and dying
next year, and the size of roots is generally proportioned to the life
of plants ; except when artificial cultivation develops the root specially, as in turnips, etc. Several of the Draconidæ are parasites, and
suck the roots of other plants, and have only just enough of their own
to catch with. The Yellow Rattle is one ; it clings to the roots of
the grasses and clovers, and no cultivation will make it thrive without
them. My authority for this last fact is Grant Allen ; but I have observed for myself that the Yellow Rattle has very small *white* suck-
ing roots, and no earth sticking to them. The toothworts and broom
rapes are Draconidæ, I think, and wholly parasites. Can it be that
the Red Rattle is the one member of the family that has 'proper pride,
and is self supporting' ? the others are mendicant orders. We had
what we choose to call the Dorcas flower show yesterday, and we gave,
as usual, prizes for wild flower bouquets. I tried to find out the local names of several flowers, but they all seemed to be called 'I
don't know, ma'am.' I would not allow this name to suffice for the
red poppy, and I said 'This red flower *must* be called *something*—tell
me what you call it ? ' A few of the audience answered ' Blind Eyes.'

Is it because they have to do with sleep that they are called Blind Eyes—or because they are dazzling ?"

20. I think, certainly, from the dazzling, which some-times with the poppy, scarlet geranium, and nasturtium, is more distinctly oppressive to the eye than a real excess of light.

I will certainly not include among my rescued Dracon-idæ, the parasitic Lathræa and Orobanche; and cannot yet make certain of any minor classification among those which I retain,—but, uniting Bartsia with Euphrasia, I shall have, in the main, the three divisions Digitalis, Lin-aria, Euphrasia, and probably separate the moneyworts as links with Veronica, and Rhinanthus as links with Lathræa.

And as I shall certainly be unable this summer, under the pressure of resumed work at Oxford, to spend time in any new botanical investigations, I will rather try to fulfil the promise given in the last number, to collect what little I have been able hitherto to describe or ascer-tain, respecting the higher modes of tree structure.

CHAPTER VII.

[The following chapter has been written six years. It was delayed in order to complete the promised clearer analysis of stem-structure; which, after a great deal of chopping, chipping, and peeling of my oaks and birches, came to reverently hopeless pause. What is here done may yet have some use in pointing out to younger students how they may simplify their language, and direct their thoughts, so as to attain, in due time, to reverent hope.]

1. THE most generally useful book, to myself, hitherto, in such little time as I have for reading about plants, has been Lindley's 'Ladies' Botany'; but the most rich and true I have yet found in illustration, the 'Histoire des Plantes,' * by Louis Figuier. I should like those of my readers who can afford it to buy both these books; the first named, at any rate, as I shall always refer to it for structural drawings, and on points of doubtful classification; while the second contains much general knowledge, expressed with some really human intelligence and feeling; besides some good and singularly *just* history of botanical discovery and the men who guided it. The botanists, indeed, tell me proudly, "Figuier is no author-

---

* Octavo : Paris, Hachette, 1865.

ity." But who wants authority! Is there nothing known
yet about plants, then, which can be taught to a boy or
girl, without referring them to an 'authority'?

I, for my own part, care only to gather what Figuier
can teach concerning things visible, to any boy or girl,
who live within reach of a bramble hedge, or a hawthorn
thicket, and can find authority enough for what they are
told, in the sticks of them.

2. If only *he* would, or could, tell us clearly that much;
but like other doctors, though with better meaning than
most, he has learned mainly to look at things with a
microscope,—rarely with his eyes. And I am sorry to
see, on re-reading this chapter of my own, which is little
more than an endeavour to analyze and arrange the statements contained in his second, that I have done it more
petulantly and unkindly than I ought; but I can't do all
the work over again, now,—more's the pity. I have not
looked at this chapter for a year, and shall be sixty before I know where I am;—(I find myself, instead, now,
sixty-four!)

3. But I stand at once partly corrected in this second
chapter of Figuier's, on the 'Tige,' French from the
Latin 'Tignum,' which 'authorities' say is again from
the Sanscrit, and means 'the thing hewn with an axe';
anyhow it is modern French for what we are to call the
stem (§ 12, p. 136).

"The tige," then, begins M. Louis, "is the axis of the
ascending system of a vegetable, and it is garnished at

intervals with vital knots, (eyes,) from which spring
leaves and buds, disposed in a perfectly regular order.
The root presents nothing of the kind. This character
permits us always to distinguish, in the vegetable axis,
what belongs really to the stem, and what to the
root."

4. Yes; and that is partly a new idea to me, for in
this power of *assigning their order* for the leaves, the
stem seems to take a royal or commandant character, and
cannot be merely defined as the connexion of the leaf
with the roots.

In *it* is put the spirit of determination. One cannot
fancy the little leaf, as it is born, determining the point
it will be born at: the governing stem must determine
that for it. Also the disorderliness of the root is to be
noted for a condition of its degradation, no less than its
love, and need, of Darkness.

Nor was I quite right (above, § 15, p. 139) in calling
the stem *itself* 'spiral': it is itself a straight-growing
rod, but one which, as it grows, lays the buds of future
leaves round it in a spiral order, like the bas-relief on
Trajan's column.

I go on with Figuier: the next passage is very valuable.

5. "The tige is the part of plants which, directed
into the air, supports, and *gives growing power to*, the
branches, the twigs, the leaves, and the flowers. The
form, strength, and direction of the tige depend on the

part that each plant has to play among the vast vegetable
population of our globe. Plants which need for their
life a pure and often-renewed air, are borne by a straight
tige, robust and tall. When they have need only of a
moist air, more condensed, and more rarely renewed,
when they have to creep on the ground or glide in thick-
ets, the tiges are long, flexible, and dragging. If they
are to float in the air, sustaining themselves on more robust vegetables, they are provided with flexible, slender,
and supple tiges."

6. Yes; but in that last sentence he loses hold of his
main idea, and to me the important one,—namely, the
connexion of the form of stem with the quality of the
air it requires. And that idea itself is at present vague,
though most valuable, to me. A strawberry creeps, with
a flexible stem, but requires certainly no less pure air
than a wood-fungus, which stands up straight. And in
our own hedges and woods, are the wild rose and honeysuckle signs of unwholesome air?

> " And honeysuckle loved to crawl
> Up the lone crags and ruined wall.
> I deemed such nooks the sweetest shade
> The sun in all his round surveyed."

It seems to me, in the nooks most haunted by honeysuckle
in my own wood, that the reason for its twining is a very
feminine one,—that it likes to twine; and that all these
whys and wherefores resolve themselves at last into—

what a modern philosopher, of course, cannot understand
—caprice.*

7. Farther on, Figuier, quoting St. Hilaire, tells us,
of the creepers in primitive forests,—"Some of them
resemble waving ribands, others coil themselves and describe vast spirals; they droop in festoons, they wind
hither and thither among the trees, they fling themselves
from one to another, and form masses of leaves and flowers in which the observer is often at a loss to discover on
which plant each several blossom grows."

For all this, the real reasons will be known only when
human beings become reasonable. For, except a curious
naturalist or wistful missionary, no Christian has trodden
the labyrinths of delight and decay among these garlands,
but men who had no other thought than how to cheat
their savage people out of their gold, and give them gin
and smallpox in exchange. But, so soon as true servants
of Heaven shall enter these Edens, and the Spirit of God
enter with them, another spirit will also be breathed into
the physical air; and the stinging insect, and venomous
snake, and poisonous tree, pass away before the power of
the regenerate human soul.

8. At length, on the structure of the tige, Figuier
begins his real work, thus:—

"A glance of the eye, thrown on the section of a log
of wood destined for warming, permits us to recognize

* See in the ninth chapter what I have been able, since this sentence
was written, to notice on the matter in question.

that the tige of the trees of our forests presents three
essential parts, which are, in going from within to without, the pith, the wood, and the bark. The pith, (in
French, marrow,) forms a sort of column in the centre
of the woody axis. In very thick and old stems its diameter appears very little; and it has even for a long
time been supposed that the marrow ends by disappear-
ing altogether from the stems of old trees. But it does
nothing of the sort;* and it is now ascertained, by exact
measures, that its diameter remains sensibly invariable†
from the moment when the young woody axis begins to
consolidate itself, to the epoch of its most complete development."

So far, so good; but what does he mean by the complete development of the young *woody* axis? When
does the axis become 'wooden,' and how far up the tree
does he call it an axis? If the stem divides into three
branches, which is the axis? And is the pith in the
trunk no thicker than in each branch?

9. He proceeds to tell us, "The marrow is formed by
a reunion of cells."—Yes, and so is Newgate, and so was
the Bastille. But what does it matter whether the marrow is made of a reunion of cells, or cellars, or walls, or

---

* I envy the French their generalized form of denial, ' Il n'en est
rien.'

† ' Sensiblement invariable ; ' ' unchanged, *so far as we can see,*' or
to general sense ; microscopic and minute change not being considered.

floors, or ceilings ?   I want to know what's the use of it ?
why doesn't it grow bigger with the rest of the tree ?
when *does* the tree ' consolidate itself ' ? when is it finally
consolidated ? and how can there be always marrow in it
when the weary frame of its age remains a mere scarred
tower of war with the elements, full of dust and bats ?

'He will tell you if only you go on patiently,' thinks

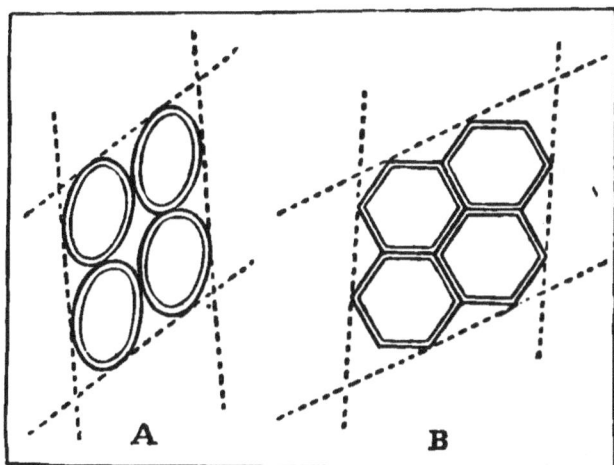

FIG. 24.

the reader.   He will not!   Once your modern botanist
gets into cells, he stays in them.   Hear how he goes on !
—"This cell is a sort of sack; this sack is completely
closed ; sometimes it is empty, sometimes it "—is full ?—
no, that would be unscientific simplicity: sometimes it
" conceals a matter in its interior."   " The marrow of
young trees, such as it is represented in Figure 24 (Figuier, Figs. 38, 39, p. 42), is nothing else"—(indeed !)—

9

" than an aggregation of cells, which, first of spherical
form, have become polyhedric by their increase and mu-
tual compression."

10. Now these figures, 38 and 39, which profess to
represent this change, show us sixteen oval cells, such as
at A, (Fig. 24) enlarged into thirteen larger, and flattish,
hexagons!—B, placed at a totally different angle.

And before I can give you the figure revised with any
available accuracy, I must know why or how the cells are
enlarged, and in what direction.

Do their walls lengthen laterally when they are empty,
or does the ' matière ' inside stuff them more out, (itself
increased from what sources?) when they are full? In
either case, during this change from circle to hexagon, is
the marrow getting thicker without getting longer? If
so, the change in the angle of the cells is intentional, and
probably is so; but the number of cells should have been
the same: and further, the term ' hexagonal ' can only be
applied to the *section* of a tubular cell, as in honeycomb,
so that the floor and ceiling of our pith cell are left un-
described.

11. Having got thus much of (partly conjectural) idea
of the mechanical structure of marrow, here follows the
solitary vital, or mortal, fact in the whole business, given
in one crushing sentence at the close :—

" The medullary tissue" (first time of using this fine
phrase for the marrow,—why can't he say marrowy tissue—' tissue moeileuse ' ?) "appears very early struck with

atony," ('atonie,' want of tone,) "above all, in its central parts." And so ends all he has to say for the present about the marrow! and it never appears to occur to him for a moment, that if indeed the noblest trees live all their lives in a state of healthy and robust paralysis, it is a distinction, hitherto unheard of, between vegetables and animals!

12. Two pages farther on, however, (p. 45,) we get more about the marrow, and of great interest,—to this effect, for I must abstract and complete here, instead of translating.

"The marrow itself is surrounded, as the centre of an electric cable is, by its guarding threads—that is to say, by a number of cords or threads coming between it and the wood, and differing from all others in the tree.

"The entire protecting cylinder composed of them has been called the 'étui,' (or needle-case,) of the marrow. But each of the cords which together form this étui, is itself composed of an almost infinitely delicate thread twisted into a screw, like the common spring of a letter-weigher or a Jack-in-the-box, but of exquisite fineness." Upon this, two pages and an elaborate figure are given to these 'trachées'—tracheas, the French call them,—and we are never told the measure of them, either in diameter or length,* and still less, the use of them!

---

* Moreover, the confusion between vertical and horizontal sections in pp. 46, 47, is completed by the misprint of vertical for horizontal in the third line of p. 43, and of horizontal for vertical in the fifth

I collect, however, in my thoughts, what I have learned thus far.

13. A tree stem, it seems, is a growing thing, cracked outside, because its skin won't stretch, paralysed inside, because its marrow won't grow, but which continues the process of its life somehow, by knitted nerves without any nervous energy in them, protected by spiral springs without any spring in them.

Stay—I am going too fast. That coiling is perhaps prepared for some kind of uncoiling; and I will try if I can't learn something about it from some other book— noticing, as I pause to think where to look, the advantage of our English tongue in its pithy Saxon word, 'pith,' separating all our ideas of vegetable structure clearly from animal; while the poor Latin and French must use the entirely inaccurate words 'medulla' and 'moelle'; all, however, concurring in their recognition of a vital power of some essential kind in this white cord of cells: "Medulla, sive illa vitalis anima est, ante se tendit, longitudinem impellens." (Pliny, 'Of the Vine,' liber X., cap. xxi.) 'Vitalis anima'—yes—*that* I accept; but 'longitudinem impellens,' I pause at; being not at all clear, yet, myself, about any impulsive power in the pith.*

---

line from bottom of p. 46; while Figure 45 is to me totally unintelligible, this being, as far as can be made out by the lettering, a section of a tree stem which has its marrow on the outside !

* " Try a bit of rhubarb" (says A, who sends me a pretty drawing

14. However, 1 take up first, and with best hope, Dr. Asa Gray, who tells me (Art. 211) that pith consists of parenchyma, 'which is at first gorged with sap,' but that many stems expand so rapidly that their pith is torn into a mere lining or into horizontal plates; and that as the stem grows older, the pith becomes dry and light, and is 'then of no farther use to the plant.' But of what use it ever was, we are not informed; and the Doctor makes us his bow, so far as the professed article on pith goes; but, farther on, I find in his account of 'Sap-wood,' (Art. 224.) that in the germinating plantlet, the sap 'ascends first through the parenchyma, especially through its central portion or pith.' Whereby we are led back to our old question, what sap is, and where it comes from, with the now superadded question, whether the young pith is a mere succulent sponge, or an active power, and constructive mechanism, nourished by the abundant sap : as Columella has it,—

"Naturali enim spiritu omne alimentum virentis quasi quædam anima, per *medullam* trunci veluti per siphonem, trahitur in summum." *

As none of these authors make any mention of a com-

---

of rhubarb pith); but as rhubarb does not grow into wood, inapplicable to our present subject; and if we descend to annual plants, rush pith is the thing to be examined.

* I am too lazy now to translate, and shall trust to the chance of some remnant, among my readers, of classical study, even in modern England.

munication between the cells of the pith, I conclude that the sap they are filled with is taken up by them, and used to construct their own thickening tissue.

15. Next, I take Balfour's 'Structural Botany,' and by his index, under the word 'Pith,' am referred to his articles 8, 72, and 75. In article 8, neither the word pith, nor any expression alluding to it, occurs.

In article 72, the stem of an outlaid tree is defined as consisting of 'pith, fibro-vascular and * woody tissue, medullary rays, bark, and epidermis.'

A more detailed statement follows, illustrated by a figure surrounded by twenty-three letters—namely, two *b* s, three *c* s, four *e* s, three *f* s, one *l*, four *m* s, three *p* s, one *r*, and two *v*s.

Eighteen or twenty minute sputters of dots may, with a good lens, be discerned to proceed from this alphabet, and to stop at various points, or lose themselves in the texture, of the represented wood. And, knowing now something of the matter beforehand, guessing a little more, and gleaning the rest with my finest glass, I achieve the elucidation of the figure, to the following extent, explicable without letters at all, by my more simple drawing, Figure 25.

16. (1) The inner circle full of little cells, diminishing in size towards the outside, represents the pith, 'very large at this period of the growth'—(the first year, we

---

* '*Or* woody tissue,' suggests A.   It is 'and' in Balfour.

are told in next page,) and 'very large'—he means in proportion to the rest of the branch. *How* large he does not say, in his text, but states, in his note, that the figure is magnified 26 diameters. I have drawn mine by the more convenient multiplier of 30, and given the real

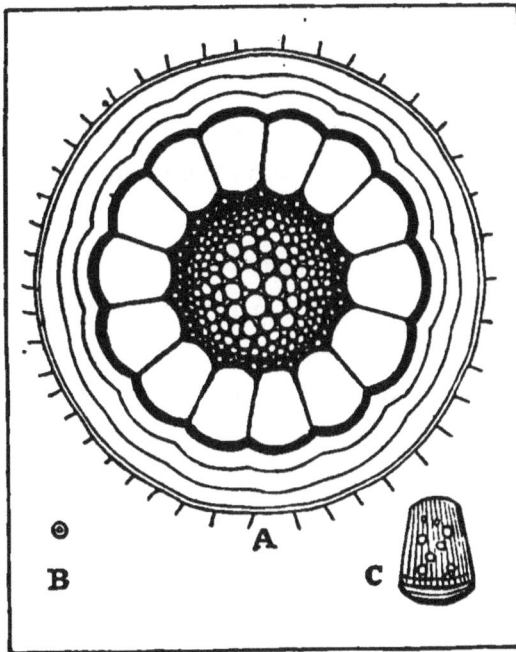

FIG. 25.

size at B, *according to Balfour :*—but without believing him to be right. I never saw a maple stem of the first year so small.

(2) The black band with white dots round the marrow, represents the marrow-sheath.

(3) From the marrow-sheath run the marrow-rays

'dividing the vascular circle into numerous compact segments.' A 'ray' cannot divide anything into a segment. Only a partition, or a knife, can do that. But we shall find presently that marrow-*rays* ought to be called marrow-*plates*, and are really mural, forming more or less continuous partitions.

(4) The compact segments 'consist of woody vessels and of porous vessels.' This is the first we have heard of woody *vessels!* He means the '*fibres* ligneux' of Figuier; and represents them in each compartment, as at C (Fig. 25), without telling us why he draws the woody vessels as radiating. They appear to radiate, indeed, when wood is sawn across, but they are really upright.

(5) A moist layer of greenish cellular tissue called the cambium layer—black in Figure 25—and he draws it in flat arches, without saying why.

(6) ⎫   Three layers of bark (called in his note Endo-
(7) ⎬ .phlœum, Mesophlœum, and Epiphlœum !) with
(8) ⎭ 'laticiferous vessels.' *

(9) Epidermis. The three layers of bark being separated by single lines, I indicate the epidermis by a double one, with a rough fringe outside, and thus we have the parts of the section clearly visible and distinct

---

* Terms not used now, but others quite as bad : Cuticle, Epidermis, Cortical layer, Periderm, Cambium, Phelloderm—six hard words for 'BARK,' says my careful annotator. "Yes ; and these new six to be changed for six newer ones next year, no doubt."

for discussion, so far as this first figure goes,—without wanting one letter of all his three and twenty!

17. But on the next page, this ingenious author gives us a new figure, which professes to represent the same order of things in a longitudinal section ; and in retracing that order sideways, instead of looking down, he not only introduces new terms, but misses one of his old layers in doing so,—thus :

His order, in explaining Figure 96, contains, as above, nine members of the tree stem.

But his order, in explaining Figure 97, contains only eight, thus :

(1) The pith.
(2) Medullary sheath. } Circles.

(3) Medullary ray = a Radius.

(4) Vascular zone, with woody *fibres* (not now vessels !)   The fibres are composed of spiral, annular, pitted, and other vessels.

(5) Inner bark or 'liber,' with layer of cambium cells.

(6) Second layer of bark, or 'cellular envelope,' with laticiferous vessels.

(7) Outer or tuberous layer of bark.

(8) Epidermis.

Doing the best I can to get at the muddle-headed gentleman's meaning, it appears, by the lettering of his Figure 97, my 25 above, that the 'liber,' number 5, contains the cambium layer in the middle of it.   The part

of the liber between the cambium and the wood is not
marked in Figure 96;—but the cambium is number 5,
and the liber outside of it is number 6,—the Endophlœum
of his note.

Having got himself into this piece of lovely confusion,
he proceeds to give a figure of the wood in the second

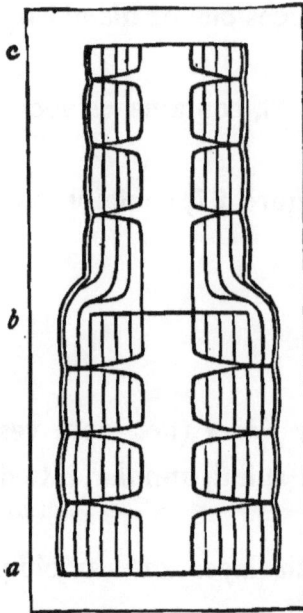

FIG. 26.

year, which I think he has bor-
rowed, without acknowledgment,
from Figuier, omitting a piece
of Figuier's woodcut which is
unexplained in Figuier's text.   I
will spare my readers the work
I have had to do, in order to get
the statements on either side
clarified : but I think they will
find, if they care to work through
the wilderness of the two authors' wits, that this which follows is the sum of what they have
effectively to tell us ; with the
collated list of the main questions
they leave unanswered—and, worse, unasked.

18. An ordinary tree branch, in transverse section,
consists essentially of three parts only,—the Pith, Wood,
and Bark.

The pith is in full animation during the first year—
that is to say, during the actual shooting of the wood.
We are left to infer that in the second year, the pith of

the then unprogressive shoot becomes collective only, not formative; and that the pith of the new shoot virtually energizes the new wood in its deposition beside the old one. Thus, let *a b*, Figure 26, be a shoot of the first year, and *b c* of the second. The pith remains of the same thickness in both, but that of the new shoot is, I suppose, chiefly active in sending down the new wood to thicken the old one, which is collected, however, and fastened by the extending pith-rays below. You see, I have given each shoot four fibres of wood for its own; then the four fibres of the upper one send out two to thicken the lower: the pith-rays, represented by the white transverse claws, catch and gather all together. Mind, I certify nothing of this to you; but if this do not happen,—let the botanists tell you what *does*.

19. Secondly. The wood, represented by these four lines, is to be always remembered as consisting of fibres and vessels; therefore it is called 'vascular,' a word which you may as well remember (though . rarely needed in familiar English), with its roots, *vas*, a vase, and *vasculum*, a little vase or phial. 'Vascule' may sometimes be allowed in botanical descriptions where 'cell' is not clear enough; thus, at present, we find our botanists calling the pith 'cellular' but the wood 'vascular,' with, I think, the implied meaning that a 'vascule,' little or large, is a long thing, and has some liquid in it, while a 'cell' is a more or less round thing, and to be supposed

empty, unless described as full. But what liquid fills
the vascules of the wood, they do not tell
us.* I assume that they absorb water, as
long as the tree lives.

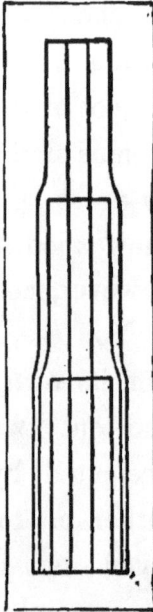

20. Wood, whether vascular or fibrous, is
however formed, in outlaid plants, first outside of the pith, and then, in shoots of the
second year, outside of the wood of the first,
and in the third year, outside of the wood of
the second; so that supposing the quantity of
wood sent down from the growing shoot
distributed on a flat plane, the structure in
the third year would be as in Figure 27.
But since the new wood is distributed all
round the stem, (in successive cords or
threads, if not at once), the increase of substance after a year or two would be untraceable, unless more
shoots than one were formed at the extremity of the
branch. Of actual bud and branch structure, I gave introductory account long since in the fifth volume of 'Modern
Painters.' † to which I would now refer the reader; but

FIG. 27.

---

* " At first the vessels are pervious and full of *fluid*, but by degrees
thickening layers are deposited, which contract their canal."—BAL-
FOUR.

† I cannot better this earlier statement, which in beginning ' Proserpina,' I intended to form a part of that work; but, as readers already
in possession of it in the original form, ought not to be burdened with
its repetition, I shall republish those chapters as a supplement, which
I trust may be soon issued.

both then, and to-day, after twenty years' further time
allowed me, I am unable to give the least explanation of
the mode in which the wood is really added to the interior stem. I cannot find, even, whether this is mainly
done in springtime, or in the summer and autumn, when
the young suckers form on the wood ; but my impression is that though all the several substances are added
annually, a little more pith going to the edges of the pith-
plates, and a little more bark to the bark, with a great
deal more wood to the wood,—there is a different or at
least successive period for each deposit, the carrying all
these elements to their places involving a fineness of basket
work or web work in the vessels, which neither microscope
nor dissecting tool can disentangle. The result on the
whole, however, is practically that we have, outside the
wood, always a mysterious 'cambium layer,' and then
some distinctions in the bark itself, of which we must
take separate notice.

21. Of Cambium, Dr. Gray's 220th article gives the
following account. "It is not a distinct substance, but
a layer of delicate new cells full of sap. The inner portion of the cambium layer is, therefore, nascent wood,
and the outer nascent bark. As the cells of this layer
multiply, the greater number lengthen vertically into
prosenchyma, or woody tissue, while some are transformed into ducts" (wood vessels?) " and others remaining as *parenchyma*, continue the medullary rays, or com-
mence new ones." Nothing is said here of the part of

the cambium which becomes bark : but at page 128, the
thin walled cells of the bark are said to be those of ordinary 'parenchyma,' and in the next page a very important passage occurs, which must have a paragraph to
itself. I close the present one with one more protest
against the entirely absurd terms ' par-enchyma,' for common cellular tissue, 'pros-enchyma,' for cellular tissue
with longer cells ;—' cambium ' for an early state of *both*,
and ' diachyma ' for a peculiar position of *one* !* while the
chemistry of all these substances is wholly neglected, and
we have no idea given us of any difference in pith, wood,
and bark, than that they are made of short or long—
young or old—cells !

22. But in Dr. Gray's 230th article comes this passage
of real value. (Italics mine—all.) " While the newer
layers of the wood abound in *crude* sap, which they convey to the leaves, those of the inner bark abound in
elaborated sap, which *they receive from the leaves*, and
convey to the *cambium* layer, or *zone of growth*. The
proper juices and peculiar products of plants are accord-
ingly found in the foliage and bark, especially the latter.
In the bark, therefore, either of the stem or root, medicinal and other principles are usually to be sought, rather

---

* " ' Diachyma ' is parenchyma in the middle of a leaf !" (Balfour,
Art. 137.) Henceforward, if I ever make botanical quotations. I shall
always call parenchyma, By-tis ; prosenchyma, To-tis ; and diachyma,
Through-tis, short for By-tissue, To-tissue, and Through-tissue—then
the student will see what all this modern wisdom comes to !

than in the wood. Nevertheless, as the wood is kept in connection with the bark by the medullary rays, many products which probably originate in the former are deposited in the wood."

23. Now, at last, I see my way to useful summary of the whole, which I had better give in a separate chapter : and will try in future to do the preliminary work of elaboration of the sap from my authorities, above shown, in its process, to the reader, without making so much fuss about it. But, I think in this case, it was desirable that the floods of pros-, par-, peri-, dia-, and circumlocution, through which one has to wade towards any emergent crag of fact in modern scientific books, should for once be seen in the wasteful tide of them ; that so I might finally pray the younger students who feel, or remember, their disastrous sway, to cure themselves for ever of the fatal habit of imagining that they know more of anything after naming it unintelligibly, and thinking about it impudently, than they did by loving sight of its nameless being, and in wise confession of its boundless mystery.

———

In re-reading the text of this number I find a few errata, noted below, and can besides secure my young readers of some things left doubtful, as, for instance, in their acceptance of the word ' Monacha,' for the flower described in the sixth chapter. I have used it now habitually too long to part with it myself, and I think it will be found serviceable and pleasurable by others. Neither shall I now change the position of the Draconidæ, as suggested at p. 118, but keep all as first planned. See among other reasons for doing so the letter quoted in p. 121.

I also add to the plate originally prepared for this number, one showing the effect of Veronica officinalis in decoration of foreground, merely by its green leaves; see the paragraphs 1 and 5 of Chapter VI. I have not represented the fine serration of the leaves, as they are quite invisible from standing height: the book should be laid on the floor and looked down on, without stooping, to see the effect intended. And so I gladly close this long-lagging number, hoping never to write such a tiresome chapter as this again, or to make so long a pause between any readable one and its sequence.

p. 105, l. 1, for ‘love’ read ‘be loved.’

p. 105, l. 3, put a semicolon, instead of comma, after ‘it.’

p. 113, l. 9 from bottom, put ‘calf’s muzzle’ in inverted commas.

p. 115, ‘never appearing in clusters’; I meant, in close masses. It forms exquisite little rosy crowds, on ground that it likes.

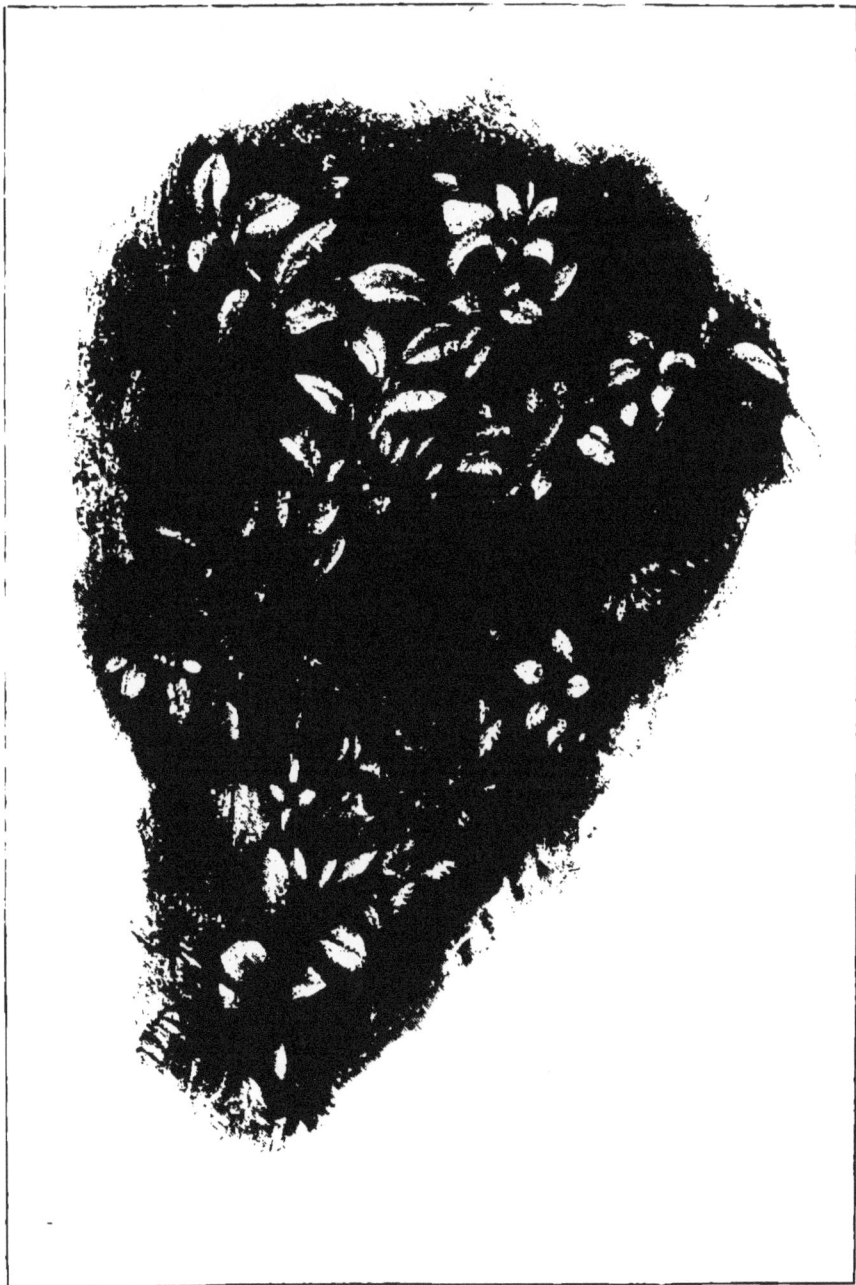

XIII.

VERONICA OFFICINALIS.

Leafage in Foreground Effect.

www.ingramcontent.com/pod-product-compliance
Lightning Source LLC
Chambersburg PA
CBHW021813190326
41518CB00007B/572